Social Media Magisterium

A No-Nonsense Guide to the Proper Use of Media

Shaun A. McAfee

ENROUTE
Make the time

En Route Books and Media, LLC
5705 Rhodes Avenue
St. Louis, MO 63109

Cover credit: TJ Burdick

Library of Congress Control Number: 2018947912

ISBN-13: 978-0-9998814-9-1
ISBN-10: 0-9998814-9-3

This book is dedicated to
Dr. Sebastian Mahfood, O.P.

Thank you for your friendship and mentoring.

CONTENTS

FOREWORD

A conversation between Jesus and a scholar of the law is recorded in Luke's gospel as follows:

> 25 There was a scholar of the law who stood up to test him and said, "Teacher, what must I do to inherit eternal life?"
> 26 Jesus said to him, "What is written in the law? How do you read it?"
> 27 He said in reply, "You shall love the Lord, your God, with all your heart, with all your being, with all your strength, and with all your mind, and your neighbor as yourself."
> 28 He replied to him, "You have answered correctly; do this and you will live."[1]

With this response, Jesus establishes the two most important commandments in our Catho-

[1] Luke 10:25-28, *New American Bible, Revised Edition* (NABRE), released on March 9, 2011, and located online at http://www.usccb.org/bible/luke/10#50010025-2 .

lic faith. He underscores that importance by explicitly stating, as recorded in Matthew's gospel, that "[t]he whole law and the prophets depend on these two commandments" (Matt. 22:40).

So important, in fact, are these commandments that the Second Vatican Council promulgated its first two documents, *Sacrosanctum Concilium*, which is the Constitution on the Sacred Liturgy, and *Inter Mirifica*, which is the Decree on the Media of Social Communications, on the same day of December 4, 1963. In doing this, the Council Fathers opened their conciliar work to the world by giving it in one stroke instruction on how to talk to God and how to talk to man.

This book by Shaun McAfee focuses on the way in which we Catholics are called to approach that second greatest commandment through the Church's instruction on our use of communicative media. Such a message is highly important in light of the pronouncement in 1 John 4:20-21,

> If anyone says, "I love God," but hates his brother, he is a liar; for whoever does not love a brother whom he has seen cannot love God whom he has not seen. This is the commandment we have from him: whoever loves God must also love his brother.

We find, then, through the scriptures and through the model of Vatican II that by addressing the second greatest commandment, we also address the first, and that by learning how to engage better in our communication with man, we may better prepare ourselves to communicate with God.

So, what are we being called to do?

To begin, the Decree on the Media of Social Communications recognizes the gift that such media, namely "the press, movies, radio, television and the like" (*Inter Mirifica*, 1), are to humankind for the impact they have on our minds. In light of that impact, in its opening paragraphs, *Inter Mirifica* examines both the service to which these tools may be put to use and the danger they may pose if put to use contrary to God's design for them.

Our missionary mandate given to us by Christ to "go into all the world and preach the gospel to all creation" (Mark 16:15), the document argues, can be facilitated through the use of media appropriate to the mission. For that reason, it is not only "an inherent right of the Church to have at its disposal and to employ any of these media insofar as they are necessary or useful for the instruction of Christians and all its efforts for the welfare of souls," but it is also "the duty of Pastors to instruct and guide the faithful so that they, with the help of these same media, may further the salvation and perfection of themselves and of the entire human family" (*IM*, 3).

The laity, furthermore, are especially called to "instill a human and Christian spirit into these media, so that they may fully measure up to the great expectations of mankind and to God's design" (*Inter Mirifica*, 3). The use of these tools, because of the impact they have on our minds, requires Christian formation in both faith and morals, for, as Marshall McLuhan pointed out in the same year with the publication of his *Understanding Media: The Extensions of Man*, our technologies first ape us and then shape us. We want to determine, in that case, which way is more desirable for us to be shaped – toward greater virtue or greater vice?

Vice has an easy definition: those habits which are contrary to virtue. The definition of virtue is harder because it rests upon the habit of excellence of a thing, and it is of two kinds, namely intellectual and moral. Aristotle explains in the first chapter of the second book of his *Nicomachean Ethics* that "intellectual virtue in the main owes both its birth and its growth to teaching (for which reason it requires experience and time), while moral virtue comes about as a result of habit."

Concerning the latter, Aristotle adds that

"states of character arise out of like activities. This is why the activities we exhibit must be of a certain kind; it is because the states of character correspond to the differences between these. It makes no small difference, then, whether we form habits of one kind or of another from our very youth; it makes a very great difference, or rather all the difference."

Inter Mirifica takes these two virtues and breaks them down, tying the intellectual virtue to the moral virtue. The decree explains that

> in society men have a right to information, in accord with the circumstances in each case, about matters concerning individuals or the community. The proper exercise of this right demands, however, that the news itself that is communicated should always be true and complete, within the bounds of justice and charity. In addition, the manner in which the news is communicated should be proper and decent.

In both the search for and reporting of the news, therefore, "there must be full respect for the laws of morality and for the legitimate rights and dignity of the individual" (*Inter Mirifica*, 5).

The moral outlook, which is concerned with

the object we are pursuing through these media, the circumstances in which we are pursuing it and our intention for the end of the pursuit, is of great importance given that reality.

What this means for consumers of communicative media is that "they must adjust their economic, political or artistic and technical aspects so as never to oppose the common good" (*Inter Mirifica*, 11). The public authorities, the document explains, are obligated to regulate this exchange to ensure that social welfare is preserved.

The Church, for her part, confirmed her teaching of six years earlier in Pope Pius XII's encyclical *Miranda Prorsus* in 1957, the purpose of which was to "give directives and instructions with regard to both sound broadcasting and television," to "turn [the technical arts of communicative media] to the extension and furthering of benefits worthy of the name." To that end, she called in *Inter Mirifica* for all children of the Church to

> join, without delay and with the greatest effort in a common work to make effective use of the media of social communication in various apostolic endeavors, as circumstances and conditions demand. They

> should anticipate harmful developments, especially in regions where more urgent efforts to advance morality and religion are needed. (13)

The methods are easily explained, and we can see the earlier form of them in *Miranda Prorsus*. First, *Inter Mirifica* prescribes, "a good press should be fostered" (14), and by 'good,' the Church calls for a Catholic press that will "disseminate and properly explain news concerning the life of the Church" (14). The obligation here is not just to the producers of press media, but also the consumers among the faithful, who "ought to be advised of the necessity both to spread and read the Catholic press to formulate Christian judgments for themselves on all events" (14).

Next, good films, radio and television programs suitable for families and that convey Church teachings should be promoted, and this requires the prompt appointment to positions within the media of Catholic clergy and laity who have skills commensurate with this apostolate and the "technical, doctrinal and moral training" of the laity, especially in the social teaching of the Church (15).

A key element of this educational process moves from the families, which are the first schools, into the parish churches, which are the second. Section sixteen reads in full,

> Since the proper use of the media of social communications which are available to audiences of different cultural backgrounds and ages, calls for instruction proper to their needs, programs which are suitable for the purpose—especially where they are designed for young people—should be encouraged, increased in numbers and organized according to Christian moral principles. This should be done in Catholic schools at every level, in seminaries and in lay apostolate groups. To speed this along catechetical manuals should present and explain Catholic teaching and regulations on this matter.

This theme about promoting the proper use of media in seminaries is important to future bishops and pastors. It's also important to Catholic schools at every level and lay apostolate groups, future parents, and lay leaders. Once the structure has been set up with the call for leadership from all ranks, *Inter Mirifica* affirms the commitment of Catholic leadership at the top when it establishes that for the purpose "that the varied apostolates of the

Church with respect to the media of social communication may be strengthened effectively, each year in every diocese of the world, by the determination of the Bishops, there should be celebrated a day on which the faithful are instructed in their responsibilities in this regard" (18).

It is in this statement that World Social Communications Day was established,[2] and that has led to over fifty years of papal leadership as each pope has, without fail each year since 1967, addressed a letter to the world concerning the use of social media. Not only do we see the popes leading from the front, but they are calling for the leadership of the bishops who should themselves invite the laity "to pray and contribute funds for this cause [which] are to be expended exclusively on the promotion, maintenance and development of institutes and undertakings of the Church in this area, according to the needs of the whole Catholic world" (18) and "to watch over such works and undertakings in their own dioceses, to promote them and, as far as the public apostolate is concerned, to guide them, not excluding those that are under the direction of the exempt religious" (20).

---

[2] The name was changed to World Communications Day beginning with the first letter from Pope John Paul II in 1979.

The World Communications Day messages both promote the positive production and consumption of media and admonish against the negative, distinguishing between the two in terms of that which promotes a Christian life and that which detracts from it. One important thing that we can draw from *Inter Mirifica* and this series of letters is that every Catholic is responsible for understanding how media affect our lives, for good or for ill, and that we are to cultivate the virtue of using media appropriately for the missionary mandate of bringing into the lives of others the light of faith and the message of Christ.

Dr. Sebastian Mahfood, OP
Professor of Interdisciplinary Studies
Holy Apostles, Cromwell, CT

Introduction

In 2013, a small Catholic parish by the name of St. Vincent de Paul on South Capitol Street in Washington, D.C., noticed a big opportunity. Though they were a small and ordinary parish, they were on the commute for thousands of fans of the MLB Nationals team, known to fans as the "Nats". As thousands of Nats fans walked by, Fr. Andrew Royals, pastor of the parish, employed what he called a "no-brainer" idea to have a Mass right before the game. The first season, the pastor reported, the parish gained about five visitors for the Mass before the game.

The next year, though, his Mass idea went viral when the parish decided to give social media a try, creating, with the help of the Catholic web-service provider named eCatholic,[3] a Facebook page and a Twitter account, branding this single Mass with the hashtag #NatsMass. That idea became a phenomenon. More and more Nats fans flocked to St. Vincent de Paul parish for Mass, which soon led Fr. Royals to realize the parish might not have the capacity in future seasons, because of its popularity, to hold just one Mass before a game. The story gained momentum and was soon talked about all over the Internet from The Washington Post to the fans themselves on their favorite apps.

The success of this campaign might seem remote and even fortuitous with the ready availability of an audience right outside its doors, but really, they're just like every other ordinary parish out there—they just figured out who their audience was. This venture should be the standard case-study for braving the world of social media in order to gain awareness and attraction to parishes. It doesn't have to be limited to Mass or even parishes. No, this sort of audience development can be put to use to gain interest and awareness for vocations, to communicate and entertain youth groups, or to

---

3 Original story from Dan Steinberg at *The Washington Post*. Story details from Joe Garcia of eCatholic, who helped St. Vincent de Paul Parish develop their website and social media to start #NatsMass campaign. See www.ecatholicwebsites.com to learn more.

simply tweet and post the already established Reconciliation times at your parish.

This pastor readily took the wisdom of Sirach to use, *"Speak concisely, say much in few words; be as one who knows and yet holds his tongue"* (32:8). If there were a verse to champion the new media, this would be it.

The fact is, this pastor took a lesson in language and put it to use. That is, he realized that communication and media is a sort of language these days, a fundamental means of disseminating information. In this case, using the branded hashtag, he was speaking the language of people already on-the-go, giving them a quick and pithy way to casually invite them to participate in something sacred.

Don't you wish you got into social media sooner when thinking of stories like this? The idea is there, but you might feel like you didn't act sooner. Which is one reason you might be reading this book, but either way, I'm telling you: it's never too late! Parishes, apostolates, activist groups, Catholic colleges, and schools all need to embark on this journey and get started before they're too late to the show and find that they cannot speak the language of their youth.

The hashtag acted as a message within a whole different language to the fans passing by, as pointed out. The new media gives us the possibility of harnessing this language. There was a point in time when mankind was united in one language. At this time, man became so powerful with their common voice that they thought they did not need God in their daily concerns. This led the people of the world to build the Tower of Babel. This tower was much more than a physical structure. It was a web of communication, intelligence, and thought. So much did this tower grow, so high, that in leaving God out of the picture, mankind sought its own destruction. Within an instant, the peoples became confused and separated. They became enemies of each other, the exact opposite of what they had once been. They created their own malicious software and content, they became infected, and they crashed.

Social media is like mankind's second chance. This new media is the single language and location that unites mankind in communication, intelligence, and thought. Though this is a digital, or virtual, continent for mankind, as it has been called, it is real. Our actual tweets, emails, blog posts, and memes are not mere figments of an unreal dimension: they are very real indeed. As real as the first words we blabber to our parents, and as effective and influential as the best of theses.

The reality of these communications, intelligentsia, and thoughts brings to us the reality that disaster is possible, just as it was in Babel. Disaster is possible because we are fragile creatures. Like the hard and soft devices that the media is built upon, we, too, are fragile hardware and software. We, too, are susceptible to viral attacks from our enemies. One attack without a good program and firewall and we can slow down, shut down, and render information and voices useless.

Even though social media can bring about our demise and destruction through our abuse of it, it can be redeemed; it can be used to preach and to spread joy, hope, love, and so much more. God doesn't rebuke these things! We can't blame God for Babel. God wishes for us to take our intellects to the maximum, and exercising these modes and media is part of that mission. In order to best use these media, we must further discover the way to renew the technology and use it rightly for the advancement of the kingdom of God.

In this book I will offer many insights into the different ways the new media—all media for that matter—can be utilized toward this end. Getting to that end requires a proper appreciation and understanding of what the Magisterium, Holy Mother Church, has to say about the media. And she has much to say. In this book, we're going to open the flood gates on what the Church has been saying for decades on the subject. By studying this, hopefully after reading this book, you or your group, organization, or even parish will be equipped to wield the power of the new media.

Chapter 1

The Magisterium

Before we begin to explore the Church's teaching on the use of Social Media, we must first understand the Church's teaching on media in general. First, we'll orient ourselves to those documents that the Magisterium has promulgated which concern media. I will also include some that discuss communication, as it is the primary role of the new media. In this chapter we'll explore the major media-related documents, their basic content, and how they can be used today from everyone from bloggers to Twitter surfers.

Vigilanti Cura

In 1936, Pope Pius XI wrote this encyclical out of a concern for the "lamentable" development of the use of cinema and the portrayal of sin and vice. The concern of the Pope was the observed use of motion pictures to an undesirable end for Christianity and mankind, that it had been used in such a way that its effect was a lowering of the moral standards of its spectators.

Though the encyclical is pointed at those who create motion picture media, the document is important to all Catholics because almost all of us consume motion picture media at some level whether for entertainment or for education. Rightfully, the document points out that it is the most popular form of media, being consumed by millions in developed and undeveloped countries. Not only the most popular, it is the most "potent" of the media arts, says the document. Quite literally, these films have the power to shape the thoughts and worldviews of generations.

Because of the popularity and the effect they have on society and culture, Pius XI indicated that the media must be "elevated," that it must be used to promote truth, virtue, knowledge of history, and justice in the world.

Multiple times, the encyclical mentions an annual pledge that the bishops of the Church have urged all to take, a pledge to "stay away" from the motion pictures and plays which offend truth and morality. I cannot help but recall the words of St. Paul to the Corinthians, commanding them to "shun immorality," literally saying to "run from" what causes us to sin.[4] We should all feel called to renew this pledge in secret, to promise not to watch a movie or stare at an image that causes us to sin. We could name many examples in 21st century television and online, so let's flee from these and seek the media that is proper to human nature.

Vigilanti Cura is important because it first outlined what effects the media have on mankind when abused and/or misused and offers a universal solution for the Church to use the media for the betterment of mankind. The central message of *Vigilanti Cura* is *responsibility*, a message that makes it even now a very engaging and relevant document when we apply it to today's media. *Vigilanti Cura* says,

> It is urgently necessary that all progress made, by God's favor, in human learning and technology actually contribute to God's glory, the salvation of souls, and the spread of the Kingdom of Jesus Christ, in such wise that we may all, as the Church bids us pray, 'so pass through the things of time that we

---

[4] 1 Corinthians 6:18.

may not lose those things that are eternal.[5]

I recommend you scan through this document along with the other documents discussed below prior to your continuing past this chapter.

Miranda Prorsus

Many consider *Miranda Prorsus* to be the first modern document of the Magisterium to concern the use of media. That's likely because the document was the first to recognize a broad scope of media arts to include radio, television, cinema, and further innovations.

Written in 1957, this was an encyclical of Pius XII. The famous element Pius XII includes in the document is the reverence for technology as "gifts from God,"[6] breaking down the perceived distrust that many have grown to have in communication innovations, a perception which many today still share. No doubt, many of us know several who have their doubts about the fruitfulness of the new media.

---

[5] Pope Pius XI, *Vigilanti Cura*, 28, available online at http://w2.vatican.va/content/pius-xi/en/encyclicals/documents/hf_p-xi_enc_29061936_vigilanti-cura.html .

[6] Pope Pius XII, *Miranda Prorsus*, 1, available online at http://w2.vatican.va/content/pius-xii/en/encyclicals/documents/hf_p-xii_enc_08091957_miranda-prorsus.html.

Miranda Prorsus contains vivid language about the interest, purpose, and use of media for the faithful. The document declares that the primary interest of adopting the new media is to fulfill the Church's mission to preach Christ to all nations.[7] The Church also has a vested interest in the development of culture and influence in society. Pius XII wished for this encyclical to be a more detailed complement to *Vigilanti Cura* promulgated by his predecessor. With this, Pius XII pronounces the concerns of the new media, and details the norms and instructions pertaining to the use of television and radio, in addition to motion pictures.[8]

Pius XII tells us that the media should be used to benefit three main functions: announcing the news, educating, and entertaining.[9] The concern Pius XII has about the new media of communication is the potential to produce and view content that carries with it grave danger to Christian life and morals. The proper use of media are those actions that promote and perfect human life and morals. All media-related actions that serve these functions serve the teachings of God and His Son, Jesus

[7] *Miranda Prorsus*, 5.
[8] *Miranda Prorsus*, 19.
[9] *Miranda Prorsus*, 47.

Christ.[10]

Inter Mirifica

Promulgated on December 4th, 1963, *Inter Mirifica* was, with *Sacrosanctum Concilium*, one of the first documents issued by the Second Vatican Council. *Inter Mirifica* is the mortar of all media-related teachings of the Church because, as an ecumenical decree, it is the document from which other teachings have emanated. It is interesting that these two were released on the same day as the first documents of Vatican II. As Dr. Sebastian Mahfood, OP, notes in his foreword, one teaches us how to speak to God, and the other teaches us how to speak to man.

This is from the opening section of the Decree on the Media of Social Communication; a document with a profound impact on media in the Church.

> Among the wonderful technological discoveries which men of talent, especially in the present era, have made with God's help, the Church welcomes and promotes with special interest those which have a most direct relation to men's minds and which

[10] *Miranda Prorsus*, 46.

have uncovered new avenues of communicating most readily news, views and teachings of every sort. The most important of these inventions are those media which, such as the press, movies, radio, television and the like, can, of their very nature, reach and influence, not only individuals, but the very masses and the whole of human society, and thus can rightly be called the media of social communication.[11]

Because this document is so invaluable to the Church's teaching on media, and the contents of this book, we will spend a few extra pages familiarizing ourselves with it.

The contents of the document set the tone for any user of media. Not just the user; the document is concerned with the producer of media to a greater degree. Their responsibility is not just the safe consumption of media, but the truthful and decent dissemination of media. As the persons with the most influence, and hence the most power, producers of media generate a great deal of expectation and have an incredible amount of responsibility. Truly, "Every one to whom much is given, of him will

---

[11] Vatican II, *Inter Mirifica*, 1, available online at http://www.vatican.va/archive/hist_councils/ii_vatican_council/documents/vat-ii_decree_19631204_inter-mirifica_en.html .

much be required" (Luke 12:48). So, what are the contents of *Inter Mirifica*? Let's quickly explore the document.

The document is separated into two chapters and 24 sections. Chapter 1 of *Inter Mirifica* is concerned with those who are affected by social media and how they must be prepared and formed to use it most effectively. The chapter also discusses basic "musts" for the user. Here are the essential messages of each section in the first half:

1. Among the technological discoveries which men have made with God's help, the Church welcomes and promotes those which have made direct relation to man's mind and have uncovered new avenues of communicating news and teaching. Because these reach all of society, they can truly be called social media.
2. These media, if properly utilized, can be of great service to mankind. But men can employ these media contrary to the plan of the creator and to their own loss.
3. The Catholic Church considers it one of its duties to announce the good news of salvation with the help of media. It is an inherent right of the Church to have at

its disposal media that are necessary for the instruction of Christians. It is the duty of pastors to instruct, and, with the help of these media, to further the plan of salvation and the perfection of themselves and the entire human family.

4. All who employ these media should be acquainted with the norms of morality and put them into practice, observing the circumstances, persons, places, and times in which they operate.
5. All parties should adopt a proper moral outlook on the use of this media. Specifically, news must be communicated true and complete.
6. With the relations between rights and the arts, and also the norms of morality, all must observe the absolute primacy of the objective moral order. For man is endowed by God with the gift of reason and some men pursue a lofty destiny.
7. The proper portrayal of moral evil can serve to bring about a deeper knowledge and study of humanity and can reveal and glorify the grand dimensions of truth and goodness.
8. In both private and public settings, every member of society must fulfill the demands of justice and charity.

9. All who make use of these media ought to avoid those things that can cause spiritual harm and lead others to danger.
10. Young people especially should take steps to accustom themselves to use self-control with the media.
11. The principal moral responsibility for the proper use of media falls on those who produce, write, and distribute the media. They must adjust their inputs so it never opposes the common good.
12. The public authority has the duty of protecting and safeguarding of freedom of information. Moreover, this authority has the duty of assisting projects that are beneficial for young people.

Notice the main audience the Council Fathers were concerned about: producers and young people. It is no coincidence that young people today are the early adopters of the new media, and the document is timeless in that regard. It doesn't take long for one to find a television show or movie from the 1950s or 1960s in which children are featured as early adopters of the new media of the day. Likewise, today, young people are adopting new forms of communication: Snapchat, Instagram, Voxer, etc. When the top social media platforms were first launched, none of their CEOs or creators were over the age of 35. *Inter Mirifica* though, warns that young people must exercise "self-control" to avoid what causes danger. This is because young persons are the most influenced by new information, as most information that comes to them is new. With all the inputs and the constant stream of influence, it is the duty of the community of believers, especially parents and pastors, to guide young people to the safe use of the new media.

Those who produce, as the Council says, have the utmost responsibility to create good and safe content. This goes far beyond the dimension of decency; it also encompasses media that can be a force of addiction to people, media that is easily penetrated by thieves, and media where users can create their own content.

It is the responsibility of creators of content *and* new media to produce safe media. Straightforward, unbiased, and decent user guidelines is a bare minimum. Sites, apps, and ecosystems must protect users from uploading and discussing obscene and hateful content. The perfect example is the Reddit, Twitter, and Youtube ecosystems, where one can not only browse for useful and interesting information and upload this same content for others to view, but can also easily access pornography, hateful speech, and terror-ridden acts and threats. The people who operate these media platforms are to be held partially responsible for not properly monitoring the accounts of those who upload this content, which has no place in the world, much less the digital continent.

Chapter 2 of *Inter Mirifica* is where the meat is! In the rest of the document, the Council Fathers focused on exactly *how* the Church must address the use of social media.

Briefly, let me diverge on a tangent. This is what is fascinating about reading a council document to me: the history-in-the-making I get to see a generation later. Being born in the mid-80s, I wasn't around for Vatican II and any of its fallout, and still, I am a convert of the post-millennium apologetics movement. So, when I entered the Catholic Church, I got a decent sense for what the Church is currently doing, but after reading Ecumenical Council documents like *Inter Mirifica*, I get to see exactly from where the Church movements within which I am living emerged. This is what makes chapter 2 exciting.

As with the first chapter, allow me to outline the essential points:

13. All the children of the Church should join in a common work to make effective the use of social media communication. Pastors, to fulfill their duty in this respect, and the laity, too, should work in this with an apostolic spirit.
14. First, a good press should be created in order to develop a fully Christian spirit in readers. Also the production of films radio and television programs should be encouraged.
15. To provide for the needs set forth here, priests, religious and lay men who are equipped with the proper skills for adapting the use of these media to the objectives of the apostolate should be appointed promptly, and they ought to be afforded technical, doctrinal, and moral training.
16. Programs that are suitable for training those of different cultural backgrounds and ages should be encouraged.
17. It is wrong for the Church to idly let the message of salvation be impeded by technical delays or expense, and this council reminds the faithful of their obligation to maintain and assist those who produce these media.

18. Each year there should be celebrated a day on which the faithful are instructed in their responsibilities with the use of social media.
19. The Supreme Pontiff should have a special office whose duties and responsibilities are dedicated to competence in the use of media.
20. It is the task of Bishops to watch over the undertaking of these media in their own dioceses and to guide them.
21. National offices should be established everywhere to see to it that the consciences of the faithful are properly instructed through these media.
22. Since the effectiveness of this media reaches beyond national boundaries of the office about which Section 21 is concerned, all media entrepreneurs and organizations should work together.

Sections 23 and 24 are appendices.

If you're like me, after reading this, you're thinking of those who responded to this great call and the organizations and domains that now exist: EWTN, Ignatius Press, New Advent, Word on Fire, En Route Books and Media, WCAT Radio, and so many more.

If you have not done so already, you should put this book down and review *Inter Mirifica.* In a short time, you will become fascinated with the foresight of the Church and be inspired by the lucidity of its contents. More than any of the other documents I have or will mention concerning the media, this book will derive its examination of the various topics from this Decree.

Communio et Progressio

Promulgated in 1971 by the Pontifical Council for Social Communications, which was itself established by *Inter Mirifica, Communio et Progressio* is not directly acknowledged as a media-related document but serves a high purpose on this book as its chief concern is proper communication. I don't think I have to explain how vital sound and ethical communication is to the use of media. Communication is the chief purpose of the media in its various roles of informing, educating, training, or entertaining.

By learning how best we should communicate corporately as a Church or individually as representatives of the Body of Christ online, we are that much better prepared to disciple others, preach the gospel, and be attractive stewards of the Kingdom. Moving forward in this book, we will integrate the teachings in *Communio et Progressio* with the teachings on media along the way.

World Communications Day Addresses

The legacy of *Inter Mirifica* lives prosperously in the modern Church. It is where the only Feast Day to emerge from the Second Vatican Council came to be. From Section 18,

> Moreover, that the varied apostolates of the Church with respect to the media of social communication may be strengthened effectively, each year in every diocese of the world, by the determination of the Bishops, there should be celebrated a day on which the faithful are instructed in their responsibilities in this regard. They should be invited to pray and contribute funds for this cause. Such funds are to be expended exclusively on the promotion, maintenance and development of institutes and undertakings of the Church in this area, according to the needs of the whole Catholic world.[12]

[12] Vatican II, *Inter Mirifica*, 18, available online at http://www.vatican.va/archive/hist_councils/ii_vatican_coun

We now know this as "World Communications Day," which has faithfully taken place each year since 1967, so for over fifty years, to educate and promote the proper use of media around the world. Each year, the Roman Pontiff delivers an address to encourage and guide the faithful in the use of media. These are highly effective documents and are never ephemeral in content, discussing a variety of topics and concerns from the service of truth, to reconciliation and evangelization, to ecumenism and the family.

Paul VI contributed twelve World Communications Day addresses. Having had a lengthy pontificate, John Paul II contributed significantly more World Communications Day addresses. From 1979 to 2005 John Paul II addressed the emerging world of media twenty-seven times. Pope Benedict XVI promulgated eight World Communications Day addresses from 2006 to 2013, and Pope Francis, the first pope with Twitter and Instagram accounts, has promulgated five World Communications Day addresses by the date of this writing. Beautifully crafted, there is a vast amount of wisdom to be drawn from these writings.

---

cil/documents/vat-ii_decree_19631204_inter-mirifica_en.html .

From Paul VI to Francis, this book will often call upon the words of the Holy Father to aid in its teaching. It will take you some time to get through them all, but if you have not done so, I invite you to read through each of them at some point. They take about 5-10 minutes to read at a time, so you can read two per day and be done within a month. It will be time well invested if you are a serious media consumer, and a must if you are a producer like a blogger, vlogger, etc.

Now that the major documents have been discussed, it is time to get started discussing the proper use of the media. Each of the aforementioned documents is available on the Vatican website for your convenience.

Chapter 2

A Re-Introduction to Social Media

What is "Social" about Social Media?

The basic identification of media is likely understood by the reader but the purpose of this book is to discuss how the *Church* views the media, old and new, so let's start there. By the way, I use the terms "new media" and "social media" interchangeably throughout this book. We identify social media by its domain, its ability to share and be discovered. Some social media is limited to status updates, others where pictures are viewed or audio is streamed, still others that are voted into history or into oblivion. The common factor is community: sharing with others the information you've discovered or created.

What makes the new media "new", though? Does the new media make the media of the days of *Inter Mirifica* "old"?

The media of yesterday is marked by its passive ability to communicate information in a one-way funnel to consumers. Viewing movies and listening to the radio are both examples of the production of broadcasting of media by one and the reception of media by one. This media is limited where producers will create, package, and send information to users who are fed that information which was seen as fit to consume.

The "new" media of today has broken down these walls, allowing users to exchange information, debate ideas, and create their own if willing and motivated to do so. The user of the new media needs but an Internet connection and a power source to relay a message to the world and can do so in "viral" speediness. Costs are low, if non-existent for some, and demand for information is at an all-time high. Whether it is a free blogging account to contribute a 1000-word article, a Facebook account to share a child's first moments with a loved one, or a Twitter account to poke fun at the latest Super Bowl commercial, the new media has made a stunning impact with its ease of access and information sharing.

That's what makes it "social". With that in mind, it doesn't make the media of yesterday obsolete, per se, because we can still apply the same values to the new and the old. We're still using the radio, television, and phone; but these media have carried messages that we can now view *and* create from anywhere. Therefore, we're called to continue applying the Church's principles and values to the new media.

No matter the mode, media become social when we embrace and interact with each other as a community. The Internet is essentially a forum, in the Roman sense, that "public pace where politics and business were transacted, where religious duties were fulfilled, where much of the social life of the city took place."[13] This digital forum is the place of social interaction today; we know it as the Internet.

The advent of the Internet, a true revolution, has truly taken social interaction with

[13] Pope John Paul II, 36th World Communications Day, 2, available online at http://w2.vatican.va/content/john-paul-ii/en/messages/communications/documents/hf_jp-ii_mes_20020122_world-communications-day.html .

media to whole new heights. This takes shape in so many ways. With new social ecosystems emerging day-by-day, there is a lot to keep up with and constant change to adapt to. What remains whether or not we use one of these or the other is our mingling, networking, growing, and learning together.

How then does the *Church* describe social media? The Council Fathers of Vatican II describe the "media of social communication" as,

> Those media which ... can, of their very nature, reach and influence, not only individuals, but the very masses and the whole of human society.[14]

By this outstandingly shrewd definition, social media is that media that can be used to help us to evangelize the world quickly and effectively. After all, to "reach and influence" masses is what we are qualified and called to do: "Go therefore and make disciples of all nations, baptizing them in the name of the Father and of the Son and of the Holy Spirit, teaching them to observe all that I have

---

[14] Vatican II, *Inter Mirifica*, 1, available online at http://www.vatican.va/archive/hist_councils/ii_vatican_council/documents/vat-ii_decree_19631204_inter-mirifica_en.html .

commanded you" (Matt. 28:19-20).

A Tradition of Excellence

Make no mistake, the Church is not new to the use of media. The Church is an early adopter and developer of every major communications innovation that has come to be invented by the best minds history has to offer but especially in our time. To make this clear is important, which is the precise reason the Vatican II Council Fathers chose to open the Decree on Social Communication I mentioned above as they did. Ultimately, as we discussed before, *Inter Mirifica* was created to continue the dialogue on how the Church must communicate to man.

> Among the wonderful technological discoveries which men of talent, especially in the present era, have made with God's help, the Church welcomes and promotes with special interest those which have a most direct relation to men's minds and which have uncovered new avenues of communicating most readily news, views and teachings of every sort.[15]

[15] Vatican II, *Inter Mirifica*, 1, available online at http://www.vatican.va/archive/hist_councils/ii_vatican_coun

Catholic inventors, developers, and adopters are many but here are a few notables.

Braille, that system of raised dots on a surface that allows for blind and visually impaired people to read, was invented in the 1820's by Louis Braille. Jean-Francois Champollion the Younger was the first producer of a translation of the Rosetta Stone. Attributed with inventing the radio and a telegraph radio system, Guglielmo Marconi is the recipient of a Nobel Prize for his work in Physics and is the father of long-distance radio transmission.

This book would be inane if it did not include the most influential Catholic figure in communications in the history of mankind, Johannes Gutenberg. His famous Bible was in the Vulgate translation and copies were known for their beautiful aesthetics. Before his invention, though, there were some prototype presses available, but hand-copies of scripture and Church documents were still the prevailing method of diffusion. His contribution to media was one of unsurpassed effect on all of the world and culture.

These are just a few highlights, but there are numerous Catholics who have been innovators

---

cil/documents/vat-ii_decree_19631204_inter-mirifica_en.html .

and great examples to the rest of us. There's one in particular that I would like to discuss, and that is Blessed James Alberione. This great priest lived from 1884 until 1971. Alberione was the fourth among a family of six children from Italy. He was ordained in 1907 and soon after received a Doctorate in Theology in 1908. Early in his ministry he received a vision that God was calling him to communicate to the world through media. On the day World War I began, he got started on making the vision he received into a reality.

He went on to found 10 religious institutes spreading to over 50 countries, including the Society of Saint Paul, The Daughters of Saint Paul, The Pious Disciples of Saint Paul, The Sisters of Jesus the Good Shepherd, and The Sisters of Mary Queen of the Apostles. These form the Pauline Family, and each of them has a special group of distinct vocations including priests, lay cooperators, single women and more. His life's work was dedicated to using the media and technology to communicate with and influence the faithful all over the world. Using the modern tools of radio, film, press, and other media, Alberione was able to reach millions with the Gospel message and leave a legacy that enabled his great work to continue.

To this day, his institutions of the Pauline

Family, especially his work founded with the help of Mother Tecla Merlo for the Daughters of Saint Paul, are strong and have a great amount of activity from the laity in the Pauline Cooperators (much more on that to come). He was revered by Pope John Paul II, being called, "The First Apostles of the New Evangelization." If he were to be a Doctor of the Church, he would be the Doctor of Media and Mass Communication. I believe in this age, we need such a title.

Father Alberione set an example worthy of remembering and vital for the correct and responsible use of social media. His example will be a reference woven into parts of this book, and anyone already or looking to become more acquainted with media would do very well to keep him as a close reference and example in their journey with media. Folks who are interested in learning more from Alberione may look at the products available from Pauline Book and Media.

There is yet another media "great" that deserves mention, and few might realize how truly Catholic this man was. The media intellectual, Marshall McLuhan, gave the world of media much to think on in terms of how we communicate and with what media we accomplish this end. He is best known for his coined, and somewhat controversial, expression "the medium is the message"—an important idea which emerged as the media began to truly become "social," a little after the mid-20th century. Along with this, and before we started calling the Internet the "digital continent," he had the idea of the "global village," the interconnectedness of the world in an electronic community. He has even been noted to have predicted the birth of the Internet thirty years before the first domain name was purchased and HTML codes were written.

His accomplishments are certainly many and his contributions to the philosophy of media are vast, but few know him for his Catholicity. He was a man of extraordinary faith and had a particular love for the Blessed Virgin. His great faith had actually played a particular role in his understanding of media, and vice versa, which we'll get into later in this chapter. His contemporaries tell of his "hard-core" conversion, his love for the early Church Fathers, and a notable influence from G.K. Chesterton's logic and jocularity in his own conversion experience. McLuhan was a media genius, and it is fitting for some of his philosophies to be included in this book. If you're interested in reading more from his work, you'll want to get a copy of *The Media and the*

Light, which is highly intellectual look at McLuhan's inspiration for his media-related philosophies.

This book would be woefully incomplete if I did not make a mention of the incredible person and work of Mother Angelica. Her fame is evidenced by her incredible reach to callers, writers, donors, and beneficiaries of her work. As she did so often on her show, she would read letters from those who heard her show on the radio, or saw her daily on television. Numerous times, she would read letters from Europe, South America, Africa, and the Far East. Still today, years after poor health removed her from the little chair she filled, *Mother Angelica Live*, her legacy is renewed. The EWTN network enjoyed astonishing growth. Within just two decades, the religious network founded by a humble nun with a vision became the largest religious media network in the world. Presently, the network is spread to 268 million listeners and viewers in more than 145 countries. From the website:

> In the early 1960s, Mother M. Angelica, a Poor Clare nun, cloistered and dedicated to the perpetual Adoration of the Most Blessed Sacrament, fulfilling a promise made to our Lord, founded Our Lady of Angels Monas-

tery in Irondale, Alabama, in the Deep South. In one of her early evangelizing efforts Mother Angelica began writing "mini-books", short teachings on a variety of religious themes. As the popularity of these "mini-books" grew, her nuns obtained a printing press and started duplicating and distributing them worldwide.

Soon, Mother began receiving requests for speaking engagements, evolving into a video series of her talks taped at a local Birmingham television station. The Poor Clare nun, who knew little of the world of technology and communication, then built her own TV studio on monastery property in Irondale, which today is home to Eternal Word Television Network.[16]

Mother Angelica was successful was several reasons, but they're very simple reasons, and they're exemplary of how we should treat the media. She was oriented to the universal call to holiness, the family, evangelism, and unity. These concepts are the very basis of this book, so she's a sure inspiration on every page.

---

[16] EWTN, *About EWTN*, available online at https://www.ewtn.com/general/index.asp.

A Common Call

Mother Church has been beckoning us to put social media to use for over sixty years. Before the call to bishops was ever made for Vatican II, before Pope John XXIII was ever elected by the College of Cardinals, the Church was promoting and pursuing the use of social media. On September 8, 1957, the feast of Our Lord's Nativity, Pope Pius XII promulgated his encyclical *Miranda Prorsus*. This document refers to the new emerging technologies (of the time) to be "gifts from God."

> Our generation takes great pride in the remarkable products of its technology, but even though these advances are the result of human talent and toil, they are still gifts of God, our Creator, from Whom all good works proceed, "for He has not only brought forth creatures; He also sustains and fosters what He has brought forth."[17]

Because the great advances in technologies are "gifts from God," we must enjoy them, put them to proper use, and cherish them. While

---

[17] Pope Pius XII, *Miranda Prorsus*, 1, available online at http://w2.vatican.va/content/pius-xii/en/encyclicals/documents/hf_p-xii_enc_08091957_miranda-prorsus.html .

there are countless blogs, articles, and other media claiming that social media makes its users less astute, truly, a gift from God cannot be so damaging to us. *Miranda Prorsus* goes on to explain the reason the Church has interest in these technologies is because it "is the duty" of Christians to announce the message of eternal salvation to men.[18]

This is an important item to note because this language sets the standard and tone for all use of media for the Church: that what we do must be for the continual purpose of announcing good will and salvation to men. This does not mean we cannot be entertaining or casual, because the technologies have allowed us to be truly communal, but we must act with these technologies with purpose and inspiration.

We can pull examples from much earlier than the days of radio and television, though. The Church has been using media in general since its foundations. We see numerous texts, some in hundreds of copies, some crisply preserved with care. The early Church understood the use of "viral" content.

What better examples of viral content do we have than Sacred Scripture itself? The words of the Apostles were not just copied on scrolls and pressed reeds. The words of Paul, Peter, and

[18] Ibid., 5.

the Evangelists were pressed into pottery and the image of the cross was formed on rings. Other Christian documents like the *Odes of Solomon, The Didache* and *The Shepherd of Hermas* gained the popularity of the masses. To this day, hundreds of these fragments and complete manuscripts have been discovered all over the ancient Christian world and in virtually every ancient language, vernacular and ecclesial. With the development of the monastic lifestyle, media became an art. Copies of Sacred Scripture were given the greatest level of craftsmanship, making up the life's work of a priest. For the laity, prayers, popular passages from Scripture, and other literature was collected and esteemed.

All of this has become rooted in the modern man as he puts hours into the production of a radio program or pours his efforts into the improvement of a script and a set for a television program. The daily blogger and podcaster are alike to these. They put forth hours of research to make the best content possible – loving their work and feeding others with spiritual food. The others who enjoy the consumption of these media are doing their part as well by hitting the share buttons and commenting, reviewing books and media, developing the conversations, and spreading good messages where they find them.

Concerning the Church's message to modern man, the use of electronic media is a topic of discussion that has not just interested one Pope or synod, but in addition to the writings of Pope Pius XII in *Miranda Prorsus*, a Decree was promulgated in Vatican II and remembered for over a half-century with annual World Communications Day addresses. Indeed, the Church is on a mission to put the best technology to the best use for the advancement of the gospel and goodwill to all mankind.

Christ, McLuhan, and the Medium

Critics of McLuhan's famous phrase, "the medium is the message," believed that he thought this was true completely, that content played absolutely no role in the information being communicated. I was one of these critics. When I read his 1964 classic *Understanding Media*, I began to compare it to the Vatican II Decree on the Media of Social Communications, *Inter Mirifica*. This document says rather explicitly that content *does* play an important role in the message communicated, naturally. It reads,

For the proper use of these media it is most

> necessary that all who employ them be acquainted with the norms of morality and conscientiously put them into practice in this area. They must look, then, to the nature of what is communicated.[19]

The Council Fathers chose to focus on discussion the importance of the nature of *what* is communicated. Further into the statement, the document refers back to the importance of good content, expressing the need for morality in the information to be broadcast and disseminated.

Again, to me, McLuhan's understanding of media was flawed, not just because his position seemed to disagree with a Magisterial document, but because it made no logical sense. I thought, *how can an .mp3 player shape my character and conscience more than the lyrics of the songs I'm listening to, or the television in my man-cave rather than the content on the dvd?*

What I didn't understand is that McLuhan held his famous phrase, "the medium is the message," to be a necessary exaggeration in order for people to realize that the medium

[19] Vatican II, *Inter Mirifica*, 4, available online at http://www.vatican.va/roman_curia/pontifical_councils/pccs/documents/rc_pc_pccs_doc_04121963_inter-mirifica_en.html .

effecttively *does* communicate a message and society at large must realize its proper effects.

McLuhan discovered and communicated something profound. His complete understanding of this point, that "the medium is the message," gave him somewhat of a theological understanding that there was something profound in this, that the only example in which the medium was the complete message is Jesus Christ. In his book, *The Medium and the Light: Reflections on Religion*, McLuhan writes,

> In Jesus Christ, there is no distance or separation between the medium and the message: it is the one case where we can say that the medium and the message are fully one and the same.[20]

This is brilliant! This truth is also iterated by the Council Fathers of Vatican II in the Dogmatic Constitution on Divine Revelation, *Dei Verbum*, "[Jesus] is both the mediator and the fullness of all revelation."[21]

Let's move forward with that on our minds

---

[20] Marshall McLuhan, *The Medium and the Light: Reflections on Religion and Media* (Berkeley, CA: Gingko Press, 1999), 103.

[21] Pope Paul VI, *Dei Verbum*, 2, available online at http://www.vatican.va/archive/hist_councils/ii_vatican_council/documents/vat-ii_const_19651118_dei-verbum_en.html .

and hearts.

Chapter 3

The Media and the Family

With the many diverse and vastly important topics possible to discuss in this book, none is immediately greater than the topic of the media and how it affects the family. Average families are challenged by the media in ways that nobody before in the history of mankind has had to endure. Evening sitcoms portraying modern families and dilemmas and education, game consoles capable of the furthest stretches of the imagination, and the possibility to chat with a parent who is deployed, the media is a very capable tool for the modern family. It is also a dangerous tool. Children can and often do stumble across pornography at an early age, witness horror and violence on shows and movies, hear foul lyrics on the radio, and learn suggestive plots in commercials and Super Bowl half-time shows.

It is not altogether evil, it is not altogether good. Being a common gathering place for families, the media must be treated with respect. This chapter is aimed at the common problems families face and how to best take up the guidance the Church offers on the subject of media, new and old.

Murder She Wrote and SNICK

Some of my earliest memories involve shows, songs, and books, but mostly shows. I remember way back in 1991 the scientist walking with April O'Neil on Teenage Mutant Ninja Turtles saying with anxiety, "Walking, talking turtles!" I remember coming home every day from the 3rd grade to my mom watching *Murder She Wrote*. We watched that *a lot* together! I remember almost every night watching my dad talk to the television set, to some guy named Alex Trebek, on Jeopardy. Then, every Saturday night, we kids watched SNICK, which was "Saturday Night Nickelodeon" – all our favorite shows for two hours. I can still rehearse the shows in order, and the theme songs in full tune. When we were old enough, my parents even gave us a buck to go rent a movie. We did all of this as a family, except for SNICK—that was for kids.

There were so many rules, though. Homework first. TV off at 10 p.m. (weekends). No PG-13 or higher rated movies. No cussing. No gore. Some of my friends thought my parents were strict. Of course, as a kid, I had little appreciation for the rules. I obeyed for the most part, but always stretched the allowances.

That was me. That was a typical 90s kid. I can't imagine what parents have to go through to keep their kids satisfied and safe in today's world. Media have the single largest effect of a child's development today, and there is no shortage of research to indicate parents must be particularly vigilant.

The Church, as we've seen, has taken a high interest in the media. The Church knows that the youth always adopt the latest technology and are, consequently, susceptible to its messages. The media are important to the development and well-being of the whole family, not just our children.

Young People

Young people are naturally equipped for the use of media. It's not a novel discovery. Young people have always been the garden from which new technologies arise. They are early adopters in two senses: first, early referring to their own lives, and early referring to the early stages of a new technology's life. Children are often quick to learn how to find a movie on Netflix, and later in life are the first to demand, test, and decide which new technologies will make it, and which ones won't.

Not only do young people use the media, but they also truly influence them. Everything from the consoles and new gear, to the content that is accessed through them, young people are already shaping the culture of tomorrow. They are as influential as they are impressionable. They require some guidance and oversight. Here's some tips on helping shape your children's involvement and well-being as they access the new media.

Any parent will watch out for three things as their child grows up: their friends, the food they consume, and the clothes they wear, among a vast number of other things. Parents rush to keep an eye on these things, but few parents truly put time into monitoring the websites, games, language, and conversations their children are having. Parents watch their children's every move in nutritional consumption, but are they doing all they can to protect them from media consumption?

What parents can do is check in on the activity of their children. I mention this in much more detail later, so I will be brief here. As parents take up this action, they must train their children to do the same for themselves. What good does it do for a person to constantly monitor and correct, but never show the peer how to make the right decisions and *why* these decisions are correct? Therefore, parents, teach your kids how to make the right decisions, show them the consequences for their actions, and, in some cases, help them understand their sins. To the kids, teens, and young adults reading this book: listen to your parents. Every time they correct you, even when you think you know better or have some bit of information that makes their instruction and discipline wrong, listen. Your parents have made almost every mistake you're about to, and it's best to listen to them and obey, even when it's not easy. Children, if you truly wish to be Godly, you must first obey those whom he has naturally placed over you for care, nurture, discipline, and love.

Young people need to watch out for a few things in particular, things which might otherwise lead to an abusive, fraudulent, and corrosive practice with the media. Young people must watch out for four things: content, programs, messages, and duration.

Content is the actual pieces of media being consumed. Like a box of tools, when they are opened and observed, the contents can be used for good, ill, or idle. We want good content. Good content is wholesome, promotional of Christian values, educational, and edifying to the human soul. Reading, listening to, and watching media of poor taste, crudity, and violence is not good content.

Programs must be carefully chosen. We all enjoy a good sitcom, but should it promote gossip? Does the new book your child is reading series contain erotica? Does the music they listen to promote violence and/or rage? Are you sure they are not watching pornography? These are tough questions. When I was fourteen I read *The Expendables*, a war novel about the true battle of la Drang in Vietnam by Leonard B. Scott. I loved war stories but around the middle of the book there was a serious love scene. It was erotic. I didn't think much of it at the time, but I went back to it, again and again. I couldn't stop thinking about this scene of forbiddingness and sensuality. Looking back, it was a title my *school teacher* should have re-read before she recommended to me—it set a blaze in me seeking other literary erotica through my formative years. Not good. There is a line we all draw, a line which continually gets pushed back as we draw closer to it. But the line must be drawn and kept. Parents, draw the line and observe it yourself. You must be an example that is in agreement with the standard you place on your children. And even if it's a book, you really need to monitor your children's consumption habits.

Next, the messages your children will receive are of high importance. What is the agenda of their newest fad? The media is a cesspool for progressive thoughts to enter unsuspectingly into the minds of young people. I would invite parents to watch these shows before and during their children's viewing time. Parents might have to pause from their media time, but there is nothing more worthwhile than the proper formation of a child.

The duration your children are able to consume media and their media devices must be thoughtfully considered. We always say with frustration, "kids these days . . ." That saying is not going away. Kids these days are becoming more and more distracted, more attached, and more like zombies than the ones on AMC. Get out of the house and go camping, hiking, or just play a sport with your child. The interaction they receive will be foundational to how they treat others, respond to real-life stress, and choose to participate in their own communities someday.

Concerning the media, we often read and discuss much of what young people are viewing, good or bad, but we rarely discuss their involvement with the production of media. That is certainly a changing dialogue in today's world. Our parents and their parents had little choice concerning what they were able to listen to on the radio, what was available on TV, with fewer channels, mind you, and no control over what the newspapers or other media produced. Today is a world different from then. Today, youth, from adolescence to teenagers and young adults, are actively producing some of the most viewed content ever, and also the most profitable. Of course, this book will tell you to help guide young people in what media is produced for it to be of goodwill, positive, and promoting of life and liberty. But what *really* must be stressed to young people and their parents is the encouragement that must projected in the first place.

Reader, if you are a millennial, and parents, if your children are of this generation, realize the incredible gift of our time: the power to reach billions with a few clicks, taps, or swipes. Anyone can create a YouTube channel, blog, vlog, or other media platform. If those in the secular world are the only ones literate in the new media, our religious views will never reach their ears. A lot of people are *still* against the use of the new media. This simply cannot be if we expect to make an impact on the next generation. If we expect to reach out and catechize the next generation, we must speak their language on their chosen media.

Consider carefully the words of Pope St. John Paul II in the 19th World Communications Day address of 1985:

> According to the right to information which every person has, communication must always respond in its content to the truth, and out of respect for the justice and charity it ought to be complete. With all the greater reason, this holds true when the communication is addressed to young people, who are in the position of opening themselves to the experiences of life.[22]

---

[22] John Paul II, 19th World Communications Day, 2, available online at http://w2.vatican.va/content/john-paul-ii/en/messages/communications/documents/hf_jp-ii_mes_15041985_world-communications-day.html .

Parents

In the opening of this chapter I told about my earliest memories involving the "tube." It's true, but they're definitely not the *only* memories. Most of my fondest memories include the times we went fishing, to baseball games, camping, hiking, and seeing Old Faithful in Yellowstone National Park. I loved being a kid! And guess what? My parents never bought us a Super Nintendo, an N64, a PlayStation, or a Genesis. The only way we got those game machines was through chores, extra chores, Christmas money, and savings. Even then, we never played Mortal Kombat.

We watched a lot of Jeopardy, and I had to sit patiently for my mom's show to come on—Star Trek. We never watched the TV during dinner, and we always did chores first. My childhood was not deprived; rather, it was enriched by the choices of my parents. But my parents were far from perfect, as was I. Later on in my teen years, I figured out how to work the world of the Internet, leading to smut and then pornography. I also figured out how to record violent movies to watch when my parents weren't around.

My parents did a good job raising us, I believe, and I'll outline some of those techniques, but there are pitfalls I and other children fall into that are avoidable through the use of appropriate methods of instilling good decision making in our children.

In the 14th World Communications Day address of 1980, Pope John Paul II notes, "It is the duty of parents to educate themselves."[23] He was speaking about the need for parents to educate themselves about the media that their children consume and use. Messages, content, duration, etc. What he was saying is the paramount lesson in this chapter: *parents must get involved*, taking up an honest and diligent interest in what children watch, listen to, read, and participate in.

Much of this involvement has already been covered in the previous section on children, but it's still important to discuss the need for parents to take up an *active* interest in their children's lives. Every parent wants what's best for his or her child, but not every parent is willing to take the time to "educate himself or herself." This section will focus a little more on that.

John Paul II also noted later in the 25th World Communications Day address of 1991

[23] John Paul II, 14th World Communications Day, available online at https://w2.vatican.va/content/john-paul-ii/en/messages/communications/documents/hf_jp-ii_mes_01051980_world-communications-day.html .

that the media can be lifeless, and life threatening, if parents don't assume responsibility.[24] Doing homework is one thing, but accepting a grade is another take that requires maturity and humility. If a child is caught doing something they have been instructed is wrong, the child is to blame, but the parents still need to evaluate themselves as well. When, if a child is caught doing something wrong because the parents failed to educate the child, the parents must assume responsibility.

Both of these are assets to parents: educating themselves and taking responsibility. First, it begins with a proper understanding of the effect the media has on the family. The media of today can bring out the very best in our families, and it can also bring about moral evil. Even when media are not bringing about moral evil, they can have negative effects.

Media can enrich. One show I rejected when I was a child was Veggie Tales. I just thought I was "cooler" than the kids who boasted about this show about a talking cucumber. When I met my wife, though, one of the first movies she had me watch was the Veggie Tales story of Jonah. I laughed so hard,

---

[24] John Paul II, 25th World Communications Day, available online at http://w2.vatican.va/content/john-paul-ii/en/messages/communications/documents/hf_jp-ii_mes_24011991_world-communications-day.html .

though it was for children, and learned so much, though it was a basic Bible story I had heard and read several times. I wondered why I never gave it a try. I knew right then and there that I wanted *my* children to be raised with it.

It's not just programs like Veggie Tales, but it's also child-friendly learning devices, video-chatting with a traveling family member, or having a phone to call home for a ride. All of these contribute to a more robust development of families, promoting support, trust, and accountability. The devices available today are more functional, accessible, and diverse—not to mention expensive—and with this can come great reward. The content available today can increase religious, scientific, and intellectual knowledge, foster identity, and nurture good character.

It can also harm. And that's an understatement. John Paul II says:

> Young people, therefore, are the first and most immediate receivers of the mass media, but they are also the most exposed to the flood of information and of images which arrive in the house by means of the media. Furthermore, it is not possible to ignore the danger of certain messages, transmitted even during the hours when

> youth listening and viewing is at its peak, slipped in under cover of ever more explicit and aggressive advertising, or introduced in shows which give the impression that the life of man is regulated only by the laws of sex and violence.[25]

The media and devices that access the media are like a warm fire: that fire can serve so many purposes. It can bring a couple together on a cool winter night. It can be the setting for a fascinating story-time. It can illuminate our imaginations. That fire can also destroy a life from the inside out. If not protected, safeguarded, and properly maintained, fires kill. Media can kill. Through promoting immoral behavior, nominalizing violence, and seeding the weeds of moral relativism, media can kill. Enough said.

Even when the media are not harmful, they can still have lingering negative effects. Media, especially the new media, have the power to create division, jealousy, envy, keep teens locked in rooms for hours, develop poor vocabulary and communication, distance parents from their children, and so much more.

---

[25] John Paul II, 19th World Communications Day, 4, available online at http://w2.vatican.va/content/john-paul-ii/en/messages/communications/documents/hf_jp-ii_mes_15041985_world-communications-day.html .

The new media, particularly, have the opportunity to bend, warp, and addict our families into bad habits and wrecked relations.

What to do? I've been negative enough, right? There are two solutions, and I've already outlined them briefly: 1) parents should inform themselves, and 2) parents must accept responsibility for the highs and the lows. Let's learn exactly how to go about both.

Research! Parents, you would never serve up a dinner without knowing two things: *what* it's made with and *who* has made it. Right? I hope so! Parents always read the labels, getting a good understanding of the ingredients and nutritional information. Then, we always know where the meal is coming from. No time in the history of man has anyone ever taken a meal and given it to their kids without getting this one down. This is even more important than the first!

How much more important is the development of a mind than the development of the stomach? So take that approach to the big screen and the small screen! Does the rating seem appropriate? Even if it does, those are based on very standardized judgments of so-called "experts." They rate based on the number of certain cuss words uttered, frequency of violence, and level of nudity and deviant behavior. They do not measure the undertones, messages, and nature of the content being delivered. *Only you, mom and dad, know what's best for your child*. Seek reviews from trusted sources. Find sources for movies and media like PluggedIn.com and well-informed voices like Steven Greydanus. Speak to your friends who've sought the same research. Most importantly, perhaps you should try the device, content, or application out for yourself. Give new apps and movies a test run. Only when you've done this should you let your kids do the same, with appropriate oversight, of course.

The research doesn't end after application. In the lab, on the job, and in the family, research ends when you evaluate your decisions. Did the show turn out to be too suggestive? Was the protagonist portrayed as a moral person, and was he/she *actually* a moral person? Is your child *ready* to have some social media account, or a media device of his or her own? There's no limit to the questions of which parents must seek answers. Be vigilant.

Once parents have educated themselves, they must educate the rest of the family—and other parents. Discuss your findings and your decisions with your children. There can be no more important function of this than to cogently communicate with your family. Even though their minds are small, children are sponges for information. And though their attitudes are at times trying, to put it lightly, they deeply desire guidance, direction, and structure.

The other side of this coin is to make sure to spread the word, to tell other parents of your discoveries and experiences. Whether it's warning or praise, you need to make sure you don't keep the information to yourself. That's what social media is all about—sharing! Don't be afraid to boldly voice your well-formed opinions and concerns publicly. The Church and the inner communities are begging for strong voices, and you're totally capable when you've shaped your mind appropriately. Facebook, Twitter, blogs, those share buttons from other's posts and Tweets, those are the 21st century's camel caravan, taking evangelization to the furthest corners of the globe.

Another approach parents must consider is the attainment and safe-keeping of their children's passwords on various media platforms. Knowing the passwords of your children adds quite an underestimated and under-discussed value to the safety of the media in your home. First, if you have a password, you can drop-in on your children's exchanges. Parents can help to advise their children on their interactions with others. Adults of Generation Y are not strangers to the experience of online bullying, threats, flirting, temptation, and manipulation. Everyone in Generation Y understands this: online interaction is worlds different from in-person interaction. Our children might not understand this immediately since they are growing up in a world immersed in online interaction. I can remember taking an hour each school day as early as the fifth grade to learn computer skills. File creation and management, typing, various database tools, and how to use the Internet to research a paper were among the lessons. This education has continued to this day—it's a continual process.

Imagine, though, that that training in the fifth grade and beyond included training in online social interaction, training that included dealings with tacit comments aimed at accusation and provocation. We cannot leave it to our schools to do this, but we have to take *action* as parents to form good online character. We can teach our kids how to reply and initiate conversation with others—it's a continual process. Our kids don't just need to learn how to use a computer as a utility; they need to learn how to use computers and the Internet as a societal function.

Checking into this and delivering the right guidance to children can ensure the right relationships are formed and nurtured as well. Parents typically pre-approve friends and get to know other families before a sleep-over, and fathers always get to know their daughter's idea of a desirable gentleman for a date. There's plenty of evidence for character in a child's profile, in the media kept on a smartphone, and the browser (and cookie) history on a browser program. Parents, use these!

In all of this, I want parents to remember one thing as they pursue this advice: always move forward with the intent of building trust. Trust is a fragile and invaluable thing. The moment a parent takes this to the level of snooping or condescension, the only reaction a child will have is distrust. That will simply inspire an interest in the child's keeping information to him- or herself.

Authority is law, but parents can be smart about this. Recall the story of God and the newly created humans in the Garden of Eden. God was open about His standards: do not eat from the tree of knowledge of good and evil (Gen. 2:17). We know what happens next. But think for a second about this. God did not password protect the whole world. When the deed was finished and sin entered the world, free will was still able to be used for good or evil. I've already suggested that you keep passwords, so I'm clearly not condemning their use.

What I'm pointing out is that parents don't have to make their children feel as though they are not capable of being trusted. Empowerment is a mighty feeling, even for children. When children feel trusted to do something right, they can soar! When children fail, help them back up. If they fail again, take appropriate measure, but don't make them feel as though they are unable to earn your trust, or incapable of making the right decisions.

There's another thing parents should be aware of and that is the ability to target privacy setting at specific individuals. What this means is that on certain social media platforms, users can set the level of content a specific individual can see on their profile. For example, if Jenny only wants her mom to be able to view status updates, but no pictures or other content, Jenny can do that. Mom might want to view her daughter's profile, but based on what Jenny has chosen, she might not see everything. This can be abused and parents should be aware of their limitation. Trust me, if kids want to outsmart their parents, they will.

A product you might have heard of but

might not have tried is Covenant Eyes. I cannot recommend this program enough. The idea is simple: accountability. How does it work? Covenant Eyes is loaded onto a computer and it tracks website content and reports to an accountability partner. This person can customize what Covenant Eyes reports and at what level they receive an alert. It's perfect for children, and adults, but this is a tool that is almost a requirement for raising young adults. Matt Fradd, apologist and popular speaker on purity, has repeatedly said that based on the information he's gathered from doctors and other surveys, you have to act as though you're assuming your child is or will use porn. There's just no more caution you can take and this is certainly not an overblown bit of advice. Covenant Eyes is low-cost, but the cost of not using it can be catastrophic.

Parents must take one thing from this chapter: be the example. Children will always grow up to act like those who had the greatest influence on them. If you want your children to be virtuous, if you truly want your family to thrive, you cannot just command it, you must live up to this expectation yourself. It's simple advice, but I understand how difficult it can be. You will fall short of perfection, but your children will see your endurance and faith. Be a good example and they will never ever forget it.

Couples

The 21st century couple has much to think about when it comes to media consumption. The first few years of my marriage, media played little part in our time together, and for me, little in my own life. I was sunk in school, earning four degrees in a row, within the span of 8 years. During this time, we had little money, no cable, and were largely unaware of or uninterested in the current TV shows. We were obedient Packers fans and fans of college hockey, but that's pretty much it.

As time went on, I became a better student, buying a lot of time for myself with more efficient study patterns. Even when I picked up new activities and hobbies, I was able to quickly adapt and free up some time for fun. Sooner or later my wife and I got into a lot of shows together. We binged on everything from *Gilmore Girls* to *Star Trek: The Next Generation*, *Friends*, and more.

Sometime into this TV binging, I began to consider what sort of content was truly appropriate for my eyes. We watched *Friends*, but I could not stand seeing the nipple scenes Rachael and Monica were popular for, or the crude jokes in the script. I told my wife I didn't want to watch it. She understood. Time passed, and with the advancement of other shows, where crude humor was a little more subtle, I began to wear down this decision, this standard I wanted to uphold.

I took a business trip sometime in 2011, was gone for a month, and wanted to keep myself busy with some shows. Netflix was a very simple means of accessing hundreds of movies and shows, and because I was alone, I didn't have to think about what my wife might have wanted to watch that night. The first few days I watched several WWII documentaries—a series my wife might snooze through. Then I watched some other movies, not paying attention to the ratings. I checked out *Gangs of New York*, not expecting the nude scenes. Then I came across a show about a time in western history I have always wanted to know more about: *The Tudors*. This was a popular semi-fictional series detailing the later life of King Henry VIII. Being somewhat of a Church history nerd, I really got excited. I had no clue what I was in for. Characteristic of HBO, the show's execution was high in drama and detail, and also characteristic of HBO, the low parts were filled with nudity and sex scenes. I was a bit perturbed, but I was alone and was really interested in the actual dialog and story.

Then my wife called. Oh boy.

She told me I was looking at porn. I was lost, "*What porn?*" and then she told me she saw the 'continue watching' part of our Netflix account and gave the show a look. She was not impressed. The anger and hurt I heard in her voice was built on a serious violation of the trust she had in me. Because I was so caught off guard, I even argued a little with her, trying to defend my decision. It was like fighting with my hands tied behind my back, a senseless way to take my blows and accept what I had gotten myself into.

In that short amount of time I went from condemning *Friends* to justifying the near-porn on The Tudors. I think we've all done something like this: we let our guard down and little by little, our standards corrode and eventually falter or cease to exist. That can happen to anyone, but couples need to take this seriously: decide on a moral standard for your consumption of media, and uphold it like it's your own roof.

I don't need to detail the topic of couples watching pornography together in this book. That's altogether the very worst decision a couple can make. It improves nothing whatsoever, least of all a sex life. However, the enemy is lurking, prowling, waiting to deceive couples in just the right way. He knows just how sacramental a marriage is and all he wants is the destruction of that. Many couples would see nothing wrong with the story above concerning the HBO series. I would challenge these people: what is the difference between a sex scene on a porn site and a sex scene on a TV show? The sex scenes in today's shows never further dialogue, character development, or plot advancement. They're only created for titillation and arousal. Don't be deceived; porn is porn. Put it this way: if you don't want your kids watching it, you shouldn't either.

Check out this quote from the *Catechism*:

> Pornography consists in removing real or simulated sexual acts from the intimacy of the partners, in order to display them deliberately to third parties. It offends against chastity because it perverts the conjugal act, the intimate giving of spouses to each other. It does grave injury to the dignity of its participants (actors, vendors, the public), since each one becomes an object of base pleasure and illicit profit for others. It immerses all who are involved in the illusion of a fantasy world. It is a grave offense. Civil authorities should prevent the

> production and distribution of pornographic materials. (2354)

Another thing I've taken note of as a married man consuming media is the time it takes to do all of this. Remember me telling you about all the studying I did and the little TV and movies we watched together? Well, we did have a bit of time, but when we did, since we had little money, we played board games, went for walks, out to eat, talked, and unmentionables. We really bonded during this time. Laughing at silly stuff, trying new recipes and cocktail mixes, and enjoyed simply being together. When the time came and we started binging on shows and I became heavily involved in media, those board game boxes gathered dust (in fact they couldn't even be found), and everything simple we used to do was all but foreign to us. We did things together, but there was no interaction, no involvement. We were together, but we weren't at the same time.

This is a commonly made mistake in marriages. Many couples fill their time being entertained by something other than each other. It's great to do things together: concerts, plays, movies. But a marriage will suffer if there is an over dependence on media and devices as the only source of passing time with each other.

Every couple consumes media to some extent. There are couples though who spend time *in the media*. Bloggers, videographers, site developers, etc. Personally, I can tell you that the Catholic media world can become somewhat addictive. The likes and retweets, the views, the clicks, the shares each bring a nice sense of affirmation. The feeling of accomplishment that comes with saying something that inspires and evangelizes another person is really great. Equal to the rewarding feeling is the cunning way we convince ourselves that we're doing more important work in the media than in the other time we could be spending with family, especially our spouses and children.

Becoming a good and well-read blogger or video producer, or other media-drive personality, takes a lot of time and devotion. This often takes us away from our families. I can say that I made this mistake. I have the ability to handle an enormous amount of work, but I've gotten to the point where I am not around for dinner many nights or have to cut out to get back to my tasks. This is different for everyone, but for you bloggers and media men and women who don't make a living in the

Catholic media, please, consider your family. Your family's hearing your talk to them about your faith is so much more important than your talking to the world wide web about it. Be careful not to fall into the trappings of a stat-driven life. Put your family first above all your media endeavors.

Elderly

As we've said in this chapter, the Church professes the fact that young people are the torchbearers and innovators of new technologies. It's important to realize this for the opportunity it presents to the world, but also the consideration the world must take on.

The elderly have a distinctive place in society. They're responsible for the soil we walk on, the education we and our parents were given, and so much more. They might not know how to use Skype, but they deserve our best in our integrating them into the new media.

Sacred Scripture tells us much about the honor our elders deserve.

> **Leviticus 19:32:** You shall stand up before the gray head and honor the face of an old man, and you shall fear your God: I am the Lord.

1 Timothy 5:1-2: Do not sharply rebuke an older man, but rather appeal to hum as a father, to the younger men as brothers, the older women as mothers, and the younger women as sisters, in all purity.

1 Peter 5:5: You younger men, likewise, be subject to your elders; and all of you, clothe yourselves with humility toward one another, for God is opposed to the proud, but gives grace to the humble.

1 Timothy 5:8: But if anyone does not provide for his own, and especially for those of his household, he has denied the faith and is worse than an unbeliever.

God's Word commands that we honor our elders. Yet, this honoring is as much a duty as it is a joy. Skyping with immobile grandparents so that they may see their grandchildren or great-grandchildren will never fail to bring us happiness.

The world is not all about the young and the restless. Neither is the media. Young people emit a spark and beauty, but the elderly are marked with wisdom, influence, experience, prudence, patience, and endurance. They are eager to see the reality of the classic philosophy, "leaving something better than you found it."

The novel and subsequent movie *The Notebook* is a fine example of the portrayal of virtue in old age and the worthwhile keeping of their health and support. To a young and naïve audience, my generation was captivated by this movie, giving hope for old age, contrary to the cultural meme.

My concern is that, today, many Catholics tend not to give enough attention to the elderly. Many young people would rather their role model fit in with the rest of the MTV generation, understand the Internet and pop culture, and have the charisma and looks to go with it. I get it, though. Young people, especially Catholics, must have the reassurance that this is not a religion for the elderly, but that other young people, too, get excited about vocations, religious education, and so on. I get that.

My hope is that we do not lose sight of those aging priests who have a world of advice and a generous spirit of virtue to hand on to us. We can read about them all day, or watch EWTN, but unless we spend actual time with them, we lose their influence. And it is such a rich influence. We become the amalgamation of the people to whom we are closest. Speaking from my personal experience, I've found a lot of success comes from surrounding oneself with the aging and elderly.

Okay, so how does this all pertain to the media? For the last paragraph, try to be sure that *all* of your friends are not just the ones of Facebook, et al. Your children's godparents, your grandparents, and your parents want to know you better, stay in touch, and see those cute children as well. Include them. If they need help setting up a webcam, help them. If they're asking for help with Twitter, or just don't understand the benefit of 140 characters, gently explain it to them. If they criticize it for developing a short attention span, ignore that (laugh to yourself later) and help them set it up anyway.

Young people, do not be afraid to teach them. They (some) are eager to learn. What I would absolutely avoid is the pressure of novelty. Just like the first generation exposed to the automobile, they don't see that their lifestyle and the world around them is changing. Even so, be gentle. Patience is a virtue. Technology has the power to intimidate people. If they feel they cannot use something so simple, and this simple tech is changing the world, they will retract and feel inept. It's especially that way with the new media, as if everyone around them learned a new language, and they didn't.

As we learned in the Bible verses earlier, nothing is more worthwhile than preserving the elderly. A couple more things on this topic must be in order, addressing the aging and elderly themselves. First, media outlets, please, if you want to preserve your apostolate, produce more media that truly affects a young audience. I know there are a lot of moving parts and influences, but the next generation is the pinnacle of significance in the future of our Church. Young people are still watching *The Soup, Keeping Up with the Kardashians*, and other reality series. Producers really need to focus and put more energy into reaching the younger generation. Remember that the medium is the message. Young people want this media, but you alone are the producer, writer, and host who can create media in which young people want to be involved. What I see going on is young people making media for young people. It's underdeveloped, under-funded, and underwhelming, but the billions of views coming from Catholics on YouTube and Vimeo show a high demand for high quality and truly Catholic media content. Please, make this happen.

Everyone consumes some form of media.

All of these forms must be considered carefully and monitored well, and this is of special importance to folks to use social media. The family is the building block of society—Vatican II called it the "first school." "Hence the family is the first school of the social virtues that every society needs."[26] Because this is true, families must pay exceptional attention to their proper consumption—and production—of media.

[26] *Gravissimum Educationis*, 3, available online at http://www.vatican.va/archive/hist_councils/ii_vatican_council/documents/vat-ii_decl_19651028_gravissimum-educationis_en.html .

Chapter 4

Evangelism: Spreading the Gospel and Embracing a Christian Witness

> Such an exhortation seems to us to be of capital importance, for the presentation of the Gospel is not an optional contribution to the Church. It is the duty incumbent on her by the command of the Lord Jesus, so that people can believe and be saved. This message is indeed necessary. It is unique. It cannot be replaced.[27]

Pope Paul VI's encyclical quoted above leaves no room for question: evangelism is the duty of each Christian. This duty extends to

---

[27] *Evangelii Nuntiandi*, 5, available online at http://w2.vatican.va/content/paul-vi/en/apost_exhortations/documents/hf_p-vi_exh_19751208_evangelii-nuntiandi.html .

both the more visible and organized occasions for evangelism, as well as the day-to-day interactions we have with others. Because we are ambassadors of Christ, the things for which we think, speak, serve, and suffer are thought, spoken, served, and suffered with a reflection of Christ. This is a pattern of behavior that comes with practice, and in the media, it becomes, as Paul VI put it, more incumbent, necessary, unique, and irreplaceable.

So what is evangelism? The reader probably already has a general idea of the meaning, but it's important to this discussion that it is revisited. Continuing with the words of Paul VI, let's examine the definition from *Evangelii Nuntiandi*. To start, the pontiff recognized the futility of attempting to lay a single definition on the term "evangelization."

> To evangelize: what meaning did this imperative have for Christ? It is certainly not easy to express in a complete synthesis the meaning, the context and the modes of evangelization as Jesus conceived it and put it into practice. In any case the attempt to make such a synthesis will never end.[28]

---

[28] *Evangelii Nuntiandi*, 7, available online at http://w2.vatican.va/content/paul-vi/en/apost_exhortations/documents/hf_p-vi_exh_19751208_evangelii-nuntiandi.html .

Though a single proper definition is difficult, Paul VI observed two essential aspects to the formula of evangelization: the announcement of the kingdom of Heaven, and the news of salvation. Paul VI writes,

> As an evangelizer, Christ first of all proclaims a kingdom, the kingdom of God; and this is so important that, by comparison, everything else becomes "the rest," which is "given I addition." Only the kingdom of Heaven therefore is absolute and it makes everything else relative.[29]
>
> As the kernel and center of His Good News, Christ proclaims salvation, this great gift of God which is liberation from everything that oppresses man but which is above all liberation from sin and the Evil One.[30]

The take away is that, in the work and mindset of the evangelist, which we learned is the obligation and honor of each Christian, there are two items always in communication:

---

[29] *Evangelii Nuntiandi*, 8, available online at http://w2.vatican.va/content/paul-vi/en/apost_exhortations/documents/hf_p-vi_exh_19751208_evangelii-nuntiandi.html .

[30] *Evangelii Nuntiandi*, 9, available online at http://w2.vatican.va/content/paul-vi/en/apost_exhortations/documents/hf_p-vi_exh_19751208_evangelii-nuntiandi.html .

the kingdom of Heaven and the gift of salvation.

Where Do We Start?

As important to the evangelist as the definition of evangelism is the "how" of evangelism. Of course, there are many ways to go about evangelizing, so there are many "how's," but there is certainly only one place to start. The absolute wrong way to start evangelizing others is by evangelizing others. The absolute right way to start evangelizing others is for the Church to evangelize yourself.[31]

You, me, our pastors, Pope Francis, all require the evangelization of ourselves first, before we attempt to show others. What good is it for the novice to become an instructor? St. Ignatius of Loyola, one of history's foremost evangelists, instructs us: "He who goes about to reform the world must begin with himself, or he loses his labor."[32] Evangelizing ourselves begins with intimacy with God through prayer,

---

[31] *Evangelii Nuntiandi*, 15, available online at http://w2.vatican.va/content/paul-vi/en/apost_exhortations/documents/hf_p-vi_exh_19751208_evangelii-nuntiandi.html .

[32] Kevin O'Brien, *The Ignatian Adventure: Experiencing the Spiritual Exercises of St. Ignatius in Daily Life* (Chicago: Loyola Press, 2011), 101.

then the call to holiness, and finally, the catechesis of our minds and hearts.[33]

Next, after we are evangelized, we must continue the process of conversion to Christ if we wish to remain fresh, vigorous, and strong in order to proclaim the Gospel.[34] The life of a faithful Christian requires a perpetual conversion in which we become more conformed to Christ in everything that we do. If we wish to think, speak, serve, and suffer like Christ, we must understand that it is not a simple and quick process—effective and correct evangelization takes time.

The Church must become conformed to Christ surely before she mandates others to follow her in doing the same. Jesus pointed this out to his disciples in a number of ways, but a striking occasion is when he instructed his followers to remain obedient to the Pharisees, but not to act like them:

> Then said Jesus to the crowds and to his disciples, "The scribes and the Pharisees sit on Moses' seat; so practice and observe

---

33 For a highly effective guide to this, get my book *Reform Yourself! How to Pray, Find Peace, and Grow in Faith with the Saints of the Counter-Reformation.*

34 *Evangelii Nuntiandi,* 15, available online at http://w2.vatican.va/content/paul-vi/en/apost_exhortations/documents/hf_p-vi_exh_19751208_evangelii-nuntiandi.html .

whatever they tell you, but not what they do; for they preach, but do not practice. They bind heavy burdens, hard to bear, and lay them on men's shoulders; but they themselves will not move them with their finger."[35]

When we think of this in a practical sense, it would be simple for a non-Christian to follow an evangelist into glory and happiness if it were only glory and happiness that await a Christian life. Unfortunately, we are not zapped into heaven the moment we are baptized, and our lives are not carefree until we meet Jesus in glory. The life of a Christian can demand everything, and for most Christians, this demands a high degree of endurance, self-denial, and hardship. Nobody will follow us into this life unless we can clearly demonstrate our own unending conversion. Nobody follows the hypocrite.

We cannot become evangelists until we are evangelized, and, still, the evangelized must continue to become transformed.

Three Acts of an Evangelist

We've so far engaged the interest of

[35] Matthew 23:1-4.

evangelism in the media by identifying a proper definition of evangelism and a common starting point. Our next item of interest, before we dive into any one media-specific item, is the essential acts of the evangelist. That would be, *what exactly is it that an evangelist does?* Going door-to-door, serving the poor, street preaching, and leading a Bible study are each secondary means of which primary duties and acts of an evangelist are visibly conducted.

In *Evangelii Nuntiandi*, Pope Paul VI lays out three primary elements to the work of the evangelist. The first is the primary order of bearing witness that God, revealed by His Son Jesus Christ, in the Holy Spirit, has loved the world and called men to eternal life with Him.[36] This is the message that God is the true and secret call inside all hearts. For many, this might be a brand new revelation from a status of complete pagan life and worship. The announcement of one God as St. Paul demonstrated at the Areopagus perfectly captures this:

> And they took hold of him and brought him to the Areopagus, saying, "May we know

[36] *Evangelii Nuntiandi*, 26, available online at http://w2.vatican.va/content/paul-vi/en/apost_exhortations/documents/hf_p-vi_exh_19751208_evangelii-nuntiandi.html .

> what this new teaching is which you present? For you bring some strange things to our ears; we wish to know therefore what these things mean." Now all the Athenians and the foreigners who lived there spent their time in nothing except telling or hearing something new.
>
> So Paul, standing in the middle of the Areopagus, said: "Men of Athens, I perceive that in every way you are very religious. For as I passed along, and observed the objects of your worship, I found also an altar with this inscription, 'To an unknown god.' What therefore you worship as unknown, this I proclaim to you.[37]

Their jaws must have dropped.

This passage tells us that "the foreigners who lived there spent their time in nothing except telling or hearing something new." Sounds like today's social media!

To our culture, this will come as a reminder of and an offer to believe and live in the faith of our ancestors, rejuvenating the faith within the culture as Nehemiah exhibited.[38]

---

[37] Acts 17:19-23.
[38] Nehemiah 8.

The second element pointed out by Paul VI is that evangelism shall always contain, as its center, a clear proclamation that in Jesus Christ, the Son of God made man, who died and rose from the grave, salvation is offered to all men as a free gift of God's grace and mercy.[39] This salvation, as pointed out by Paul VI, is not a temporary means of material and spiritual wealth, but an absolute, eternal, and transcendent salvation of which we may both participate and receive. The salvation offered here by the evangelist is not a better life—it is eternal life. This salvation offered through God's Son Jesus Christ has its beginning in this life and is fulfilled in eternity.

Along with the previous elements, the witness of the reality of an almighty and loving God, and the proclamation of the gospel of salvation, is the prophetic and natural calling to a world beyond the norm. The third element and act of the evangelist is the preaching of the promises made by God in the new Covenant of Jesus Christ.[40] This message is inclusive of the life one is to adopt in order to be a Christian:

---

39 *Evangelii Nuntiandi*, 27, available online at http://w2.vatican.va/content/paul-vi/en/apost_exhortations/documents/hf_p-vi_exh_19751208_evangelii-nuntiandi.html .

40 *Evangelii Nuntiandi*, 28, available online at http://w2.vatican.va/content/paul-vi/en/apost_exhortations/documents/hf_p-vi_exh_19751208_evangelii-nuntiandi.html .

acknowledging God and his love for us, giving and forgiving, prayer and fasting, participation in the sacraments, and living a life of self-denial. This lifestyle is not meant to impede and corrode our earthly life, but it is meant to be the fulfilment and perfection thereof.

Although, as we have learned, the salvation offered in the evangelist's message does not mean an end to the struggles, stress, and suffering experienced in this lifetime, but the Christian message is one that is full of faith, hope, and love. With these three, in the promises of Jesus Christ, life is certainly better in the sense of its worth, its meaning, and its enjoyment.

The main thing to take away from this is that these three acts are the three prerogatives of an evangelist. Because, as we learned, evangelism is a constant duty of all Christians, these must be the constant basis for consumption and production for all involved in the media. These three make up the general *spirit* of the behavior and judgment.

That might seem rather dull and tedious. *Does everything I do have to be directly involved with proclaiming God, announcing salvation, and preaching God's promises?*

Well, yes. And no.

Insofar as we watch programs that inspire goodness, convict our sins, and bring us closer to God, we are being evangelized. As much as we view and post righteous content online, we are promoting and proclaiming God's kingdom and His promises. When we seek wholesome entertainment, we are spreading love, joy, goodness, God. When we read books on history and watch shows about justice, we are seeking truth. God has filled the world with knowledge, and he calls us to discover a degree of amusement for our pleasure, spectacles for our wonderment, and displays of integrity to enhance our judgment. When we consume and create these things, they can be used for evangelization.

Every Tweet, status, post, comment, view, page turn, channel, disk, game, and text does not need to contain content explicit of our beliefs, but it should always be implicit of these beliefs. There are numerous ways, then, for us to evangelize. No one, single formula of practice and content should be given as a directive to readers. In addition to the three acts of the evangelist given to us by Paul VI, there are some important challenges for our time that Catholics must strive in the vineyard to address. The rest of this chapter will focus on those evangelization efforts.

Promote Holiness

Many Christians struggle to understand what God wants them to do. They lament year after year their spiritual drought of not being used in the kingdom and not knowing what God's will is for them. These Christians are looking for something *to do* when they should be looking for someone *to be*. They are looking for God's specific will for their action when they are missing is perfecting God's will for their life. For each of us, His will is plainly specific: "For this is the will of God, your sanctification: that you abstain from immorality."[41]

These are clear instructions: be holy and refrain from immoral acts. More than the souls God wants you to win, he wants your soul, and the way to give it to him is to be holy.

After this, evangelization demands that we promote holiness in others. The three acts of the evangelist involve preaching the kerygma, but Christians must not only *know* about the good news, they must be equipped to live a life that reflects the goodness of Christ. Soldiers are not just recruited—they are trained and equipped to perform.

Evangelization with social media comes

[41] 1 Thessalonians 4:3.

naturally to some, both for others it takes some courage and for still others, practice. I believe that the simplest way to promote holiness in others with social media is to share; share what you find; share what you think and believe; share what you write and what your Catholic friends write. Meaningful iconography, blog posts, news, quotes, and more are each very effective means of sharing the gospel with others online and promoting holiness. While you do this, engage others in conversation. Have courage to discuss issues on morality with Catholics and non-Catholics alike. And while you do this, keep in mind that, again, the biggest priority in all of this is to be true to the message you are preaching. Pope Benedict XVI put it this way:

> To proclaim the Gospel through the new media means not only to insert expressly religious content into different media platforms, but also to witness consistently, in one's own digital profile and in the way one communicates choices, preferences and judgements [sic] that are fully consistent with the Gospel, even when it is not spoken of specifically. Furthermore, it is also true in the digital world that a message cannot be proclaimed without a consistent witness

> on the part of the one who proclaims it. In these new circumstances and with these new forms of expression, Christians are once again called to offer a response to anyone who asks for a reason for the hope that is within them.[42]

Social Justice and Modern Abolition

Evangelizers are also abolitionists. Some people believe that religion has no place in politics, but how can it be a religion for our life if it does not extend to the politics that affect our life? The answer is simple: the premise is false. Religion has everything to do with politics because it has everything to do with worldview, justice, law-making, how we treat criminals and enemies, and everything else in between. The evangelist is an abolitionist.

Recall St. Paul's letter to Philemon. Paul did not require it, but he urged that the master let the slave Onesimus go free (Phil. 1:1-20). It was not just that the slave was a great value to Paul, but also that the spirit in which Paul spoke was

[42] Benedict XVI, 45th World Communications Day, available online at http://w2.vatican.va/content/benedict-xvi/en/messages/communications/documents/hf_ben-xvi_mes_20110124_45th-world-communications-day.html .

one of freedom and dignity for life. Paul was a missionary, but he knew the truth of the Gospel, which caused him to say something about a situation in which he had some influence and personal interest.

That's what it takes for today's media-using Catholic men and women. We have to be abolitionists. I'm so proud of what I've seen and been part of so far: young voices like Abby Johnson leading the charge for our generation's fight against abortion; men like Matt Fradd boldly dedicating his life to "kicking porn in the face;" my own wife writing and sharing the good news about Natural Family Planning (NFP), and a plethora of other young people putting their God-given talents to use to expose the crimes and change the world in which we live. These are the Modern Abolitionists. But the world needs more voices.

We can follow their example. There must be a charismatic Catholic who's worked in the criminal system, being an eyewitness to capital punishment. There's got to be a voice to speak about the elderly and the value of their lives even when infirm, or a doctor with personal knowledge about euthanasia, or a mother with a child born with a disease or special condition (I know one - Jeannie Ewing), or a successful marriage once on the brink of collapse. The possibilities stretch as far as the latest abuse to human dignity.

Following the example of these men and women means a few things, and there are a few things it doesn't mean. First, it means seeking God's will above all things, not just aimlessly going and doing anything at will. There are many who justify their actions solely because they are good, but they don't consider that out of the almost infinite good things to do in the world, God has a particular mission for our lives. Not everyone needs to be Scott Hahn. Second, it means joining and supporting them, not competing against them. Far too many people are eager to build their own platform rather than be a pillar. Third, it means gathering people with talents to assist in the apostolate, not being a one-man-show. Get plugged in, and get your story out there!

Chapter 5

Ecumenism: Sharing the Idea of Unity

What is Ecumenism? And isn't it a bad thing?

Ecumenism is the word we use for actions and dialogue that promote and enhance unity among Christians. Ecumenism is also our central aim in the conversion and salvation of the world. Indeed, in heaven, there is not a room full of Catholics, a room full of Southern Baptists, and another room full of "non-denominational" Christians. We will all occupy that place in unity and with charity as we are one Church and one Body of Christ.

However, in some circles, ecumenism is used negatively. To some, ecumenism is nothing more than the inclusive progressivism of the Christian faith. To these people, ecumenical movements are nothing more than occasions where the Christians Faith is watered down to appease all believers. That's not ecumenism. If anything, that's the opposite. Ecumenical movements seek unity, yes, but not at the expense of truth.

Unity is one of the marks of the Church and is mentioned in the Nicene Creed: "I believe in one, holy, catholic and apostolic church." The *Catechism of the Catholic Church* has much to say on unity within the Church, but paragraphs 813 and 820 serve this conversation perfectly, to define the term and to display the importance of the act.

> The Church is one because of her source: "the highest exemplar and source of this mystery is the unity, in the Trinity of Persons, of one God, the Father and the Son in the Holy Spirit. The Church is one because of her founder: for the Word made flesh, the prince of peace, reconciled all men to God by the cross . . . restoring the unity of all in one people and one body." The Church is one because of her "soul": "It is the Holy Spirit, dwelling in those who believe and pervading and ruling over the entire Church, who brings about that

wonderful communion of the faithful and joins them together so intimately in Christ that he is the principle of the Church's unity" (CCC 813).

Christ bestowed unity on his Church from the beginning. This unity, we believe, subsists in the Catholic Church as something she can never lose, and we hope that it will continue to increase until the end of time. Christ always gives his Church the gift of unity, but the Church must always pray and work to maintain, reinforce, and perfect the unity that Christ wills for her. This is why Jesus himself prayed at the hour of his Passion, and does not cease praying to his Father, for the unity of his disciples: "That they may all be one. As you, Father, are in me and I am in you, may they also be one in us . . . so that the world may know that you have sent me." The desire to recover the unity of all Christians is a gift of Christ and a call of the Holy Spirit (CCC 820).

The need for unity, though, is not a new initiative: it truly goes back to the prayer of Jesus in the garden of Gethsemane. Jesus' prayer is filled with mentions of his desire for the unity of his followers. His penultimate request to the Father reads:

> I do not pray for these only, but also for those who believe in me through their word, that they may all be one; even as thou, Father, art in me, and I in thee, that they also may be in us, so that the world may believe that thou hast sent me. The glory which thou hast given me I have given to them, that they may be one even as we are one, I in them and thou in me, that they may become perfectly one, so that the world may know that thou hast sent me and hast loved them even as thou hast loved me (John 17:20-24).

This prayer serves as a fiat in Jesus, that when we promote unity and engage in activities that contribute to unity, we contribute to four important effects:

- We glorify God.
- We emulate the unity of the Trinity.
- We help the world to believe in the love of God.
- We help the world to believe in the sending of Christ to the world.

Judging by Jesus' own prayer, unity is no

matter of relativity but is the makeup of everything we believe. It makes sense, too. Think of marriage as an analogy. As we watch our parents observe their promises of fidelity, love each other, and stay as one flesh, we are much more likely to believe in the institution of marriage as something that is good. In the same way, when the world sees Christians interacting with love and unity, we promote the deepest values of our institution and ultimately the One in Whom we have faith: God.

In the history of the Church, the message of ecumenism has been promoted from the apostolic age to the Church Fathers, through the saints and from the writings of popes. Briefly, this chapter will explore those voices.

Irenaeus of Lyons, Bishop in Gaul who learned the faith from the preaching of St. Polycarp, a disciple of St. John the Evangelist, said in the second century:

> The Church, though dispersed throughout the whole world, even to the ends of the earth, has received from the apostles and their disciples this faith.[43]

St. Cyprian of the third century wrote the following two declarations on the unity of the

[43] Irenaeus of Lyons, *Against Heresies*, 10.1.

Church:

> The spouse of Christ cannot be adulterous; she is uncorrupted and pure. She knows one home; she guards with chaste modesty the sanctity of one couch. She keeps us for God. She appoints the sons whom she has born for the kingdom.[44]

> You have written also, that on my account the Church has now a portion of herself in a state of dispersion, although the whole people of the Church are collected, and united, and joined to itself in an undivided concord: they alone have remained without, who even, if they had been within, would have had to be cast out.[45]

These are among the earliest voices to describe the Church's position on unity, a significant part of the Christian Faith that has not been lost to time. The Magisterium has eloquent voices in the words of the papal encyclicals as well. From Clement XIII to John Paul II, the popes have spoken at length about the need for ecumenism.

---

[44] Cyprian of Carthage, On the Unity of the Church *De Unitate Ecclesiae*, 6.

[45] Cyprian of Carthage, *The Epistles of Cyprian*, 66.8.

Clement XIII's first encyclical letter after becoming supreme pontiff was devoted to ecumenism. Written in 1758 and entitled *A Quo Die*, he wrote the following words of wisdom:

> You can easily see that there is a difference between the sons of light and the sons of the world. The latter disagree among themselves with various and diverse opinions, while the former, initiated into the mysteries of unity, profess the one faith of all by the mouth of one, through the head of all. Therefore, concentrate all your attention on increasing peace among the faithful. Uproars, contentions, rivalries, animosities, and dissensions should be silenced. In this way those who go by the name of Catholic can all be perfect in the same sense, in the same opinion, saying the same thing together, knowing the same thing and understanding it thoroughly. They should understand that if they want to be members of Christ, they cannot have concord with the head if they want to be in disagreement with the members. Nor can those who have not lived in fraternal love be counted as brothers by the Almighty Father.[46]

[46] Clement XIII, *A Quo Die*, 3, available online at http://www.papalencyclicals.net/clem13/c13aquod.htm .

Being ecumenical also has much to do with truth. Indeed, if we are to remain as one and promote the Church to reach unity, doctrine and sacred Tradition are not "up for grabs" or subject to negotiation. Pius IX, almost a century later in 1853, wrote his ninth encyclical letter, dedicating it to the practical means in which both the clergy and laity can contribute to ecumenical activity. Written to the Church in France, the letter exhorts many different activities, but one that he gave can be applied to our age of digital media: encouraging Catholic writers. He wrote:

> And here we cannot help but remind you of the admonitions and counsels with which, four years ago, we strongly summoned all the bishops of the Catholic world to exhort men outstanding for talent and sound doctrine to publish appropriate writings with which they might enlighten the minds of people and dissipate the darkness of creeping errors. Strive to remove this deadly pestilence of books and magazines from the faithful given into your care. At the same time encourage with all benevolence and favor those men who, animated by a Catholic spirit and educated in literature and learning, will endeavor to write books

> and publish magazines. Do this so that the Catholic doctrine is defended and spread, that the venerable rights and documents of this Holy See remain sound, that opinions and doctrines opposed to the same See and its authority may be suppressed, and that the darkness of error is banished and the minds of men illumined with the sweet light of truth. And it will be for your episcopal solicitude and love to arouse such inspired Catholic writers, so that they continue with ever greater zeal and knowledge to defend the cause of Catholic truth. You must also admonish them like a prudent father if their writings should offend Catholic teaching.[47]

Within the plethora of writings Leo XIII contributed to the Church's Magisterium can be found *Satis Cognitum*, promulgated on June 29, 1896. This document apologetically defends the Church's teaching on ecclesiology concerning unity, with stress placed primarily on the need and truth of the primacy of the Supreme Pontiff, where the Church derives her unity. In the text of *Satis Cognitum*, Leo XIII clarifies in the opening paragraph:

---

[47] Pius IX, *Inter Multiplices*, 6, available online at http://www.papalencyclicals.net/pius09/p9interm.htm .

> It is sufficiently well known unto you that no small share of Our thoughts and of Our care is devoted to Our endeavor to bring back to the fold, placed under the guardianship of Jesus Christ, the Chief Pastor of souls, sheep that have strayed. Bent upon this, We have thought it most conducive to this salutary end and purpose to describe the exemplar and, as it were, the lineaments of the Church. Amongst these the most worthy of Our chief consideration is Unity.[48]

The words of the pope could not be more efficient: it is the primary concern of the Church to bring all Christians to unity. Quoting St. Cyprian of Carthage, the encyclical, again, powerfully commands the importance of unity:

> The Church of Christ, therefore, is one and the same forever; those who leave it depart from the will and command of Christ, the Lord—leaving the path of salvation they enter on that of perdition. "Whosoever is separated from the Church is united to an adulteress. He has cut himself off from the

[48] Leo XIII, *Satis Cognitum*, 1, available online at http://w2.vatican.va/content/leo-xiii/en/encyclicals/documents/hf_l-xiii_enc_29061896_satis-cognitum.html .

> promises of the Church, and he who leaves the Church of Christ cannot arrive at the rewards of Christ. . . . He who observes not this unity observes not the law of God, holds not the faith of the Father and the Son, clings not to life and salvation."

Generations later, ecumenism was a central part of the Second Vatican Council. Actually, that's an understatement. Of course, the name "ecumenical council" signifies the meaning, but Vatican II was especially concerned with the concept of unity. The opening speech by Pope John XXIII gives a glimpse into the vision the pontiff had when he called the meeting to be:

> The Catholic Church, therefore, considers it her duty to work actively so that there may be fulfilled the great mystery of that unity, which Jesus Christ invoked with fervent prayer from His heavenly Father on the eve of His sacrifice. She rejoices in peace, knowing well that she is intimately associated with that prayer, and then exults greatly at seeing that invocation extend its efficacy with salutary fruit, even among those who are outside her fold.
>
> Indeed, if one considers well this same

> unity which Christ implored for his Church, it seems to shine, as it were, with a triple ray of beneficent supernal light: namely, the unity of Catholics among themselves, which must always be kept exemplary and most firm; the unity of prayers and ardent desires with which those Christians separated from this Apostolic See aspire to be united with us.[49]

John XXIII died just eight months into the sessions of Vatican II, but the vision was dutifully carried on by his successor, Pope Paul VI. This bold and learned pope—while the Council was still ongoing—made heads turn with his 1964 encyclical, *Ecclesiam Suam*. This was to be the first encyclical of its kind on the topic of ecumenism, and it would shape thereafter the outcome and promulgations of Vatican II.

There were many encyclicals written on the topic of ecumenism before Paul VI's as we've explored already, but Paul VI's letter was different. Each of the documents before *Ecclesiam Suam* was limited to the context of the Catholic Church exclusively, not to the

---

[49] John XXIII, *Speech for the Solemn Opening of the Ecumenical Council Vatican II*, 8, available online at https://www.catholicculture.org/culture/library/view.cfm?recnum=3233 .

Christian church at large. The direction given in these letters concerned the unity which we were to *maintain* with our Catholic family, not the relationship that needed to be *reconciled* with our non-Catholic brothers and sisters. *Ecclesiam Suam* was different because it changed the tone of ecumenical voice of the Church, and that difference can be summed up in one word: invitation.

> We readily accept the principle of stressing what we all have in common rather than what divides us. This provides a good and fruitful basis for our dialogue, and we are prepared to engage upon it with a will. We would even go further and declare our readiness to examine how we can meet the legitimate desires of our separated Christian brothers on many points of difference concerning tradition, spirituality, canon law, and worship, for it is Our dearest wish to embrace them in a perfect union of faith and charity.[50]

Because of the change in tone on ecumenism from *Ecclesiam Suam*, the docu-

---

[50] Paul VI, *Ecclesiam Suam*, 109, available online at http://w2.vatican.va/content/paul-vi/en/encyclicals/documents/hf_p-vi_enc_06081964_ecclesiam.html .

ments emanating from Vatican II were likely influenced. Vatican II promulgated *Unitatis Redintegratio,* The Decree on Ecumenism, on November 21, 1964. This docu-ment cogently details the ecumenical situation between the Catholic Church and her Protestant and Eastern Christian brethren, and it outlines the communication and functions that will facilitate more fruitful dialogue between Christians.

> We must get to know the outlook of our separated brethren. To achieve this purpose, study is of necessity required, and this must be pursued with a sense of realism and good will. Catholics, who already have a proper grounding, need to acquire a more adequate understanding of the respective doctrines of our separated brethren, their history, their spiritual and liturgical life, their religious psychology and general background. Most valuable for this purpose are meetings of the two sides-especially for discussion of theological problems-where each can treat with the other on an equal footing-provided that those who take part in them are truly competent and have the approval of the bishops. From such dialogue will emerge still more clearly what

the situation of the Catholic Church really is. In this way too the outlook of our separated brethren will be better understood, and our own belief more aptly explained.[51]

Years later, John Paul II also spoke and wrote extensively on the topic of ecumenism in his landmark encyclical *Ut Unum Sint*. In this encyclical, John Paul II outlines our proper relationship to other Christians and how exactly we can work out a better dialogue with them. John Paul II's message in *Ut Unum Sint* can be found in the second section when he states that "No [Christian] is unaware of the challenge which all [schism] poses to believers." But, he goes on, "They cannot fail to meet this challenge."[52]

We must rise up as Catholics and actively seek the unity on which these popes and early apologists insisted. After all, it is the chief mission of the Church.

---

[51] *Unitatis Redintegratio*, 9, available online at http://www.vatican.va/archive/hist_councils/ii_vatican_council/documents/vat-ii_decree_19641121_unitatis-redintegratio_en.html .

[52] John Paul II, *Ut Unum Sint*, 2, available online at http://w2.vatican.va/content/john-paul-ii/en/encyclicals/documents/hf_jp-ii_enc_25051995_ut-unum-sint.html .

Ecumenism and the Media

Ecumenism is a central theme of social communications. Really, it's what social media is all about, hence, "social." We all have different choices on email service providers, and we all seek our friends of this or that belief on Facebook, Twitter, and other ecosystems. Though we are all on different domains, different servers, and varying locations, we are still able to dialogue. We should make our dialogue a religiously ecumenical one.

It's important to go forward with the understanding that we are all on the same team in media. Statistics matter. They do (and I discuss this later in the book), but so does teamwork. When one media innovator comes up with an idea, the media community and individuals should be there to support, encourage, and improve upon the content and application of such an innovation. Bloggers, podcasters, and radio personalities must work as one body to preach Christ to the world and to glorify the Father through the Holy Spirit, not as individuals reaching for their share of the demographic pie.

Ecumenical Dialogue in Social Media

In order to accomplish this, firstly, we must be unified with one another if we ever hope to achieve ecumenical communication with other Christians. Part of the reason some people wish not to engage with certain social groups is due to the perception of that group's lacking togetherness.

What does this look like? It takes form in the daily interactions in comment threads, Tweets, radio show callers, and Facebook posts. Each of these and those related should engage topics without vitriol and rudeness, but to embrace the opinions and decisions of others so long as they are not contrary to Catholic teaching. Even then, gentleness is the best course of action (1 Pet. 3:16). Catholic media is filled with topics to discuss and it is the job of those consuming and enjoying the conversation to also 'police' the threads before they become self-destructive and corrosive to our image as a body. Most of all, it takes place in the actions of those who blog and contribute to producing other forms of media. Hosts who can handle the angry caller, writers who can handle the "troll" and producers who can harmoniously organize personnel and positive content will drive the engine of ecumenism.

This is usually a matter of individual appearance and behavior, but there is an adequate way for the community of Catholics to seek unity as well. The *Catechism of the Catholic Church* says that there are seven such means:

- A permanent renewal of the Church in greater fidelity to her vocation; such renewal is the driving-force of the movement toward unity;
- Conversion of heart as the faithful "try to live holier lives according to the Gospel"; for it is the unfaithfulness of the members to Christ's gift which causes divisions;
- Prayer in common, because "change of heart and holiness of life, along with public and private prayer for the unity of Christians, should be regarded as the soul of the whole ecumenical movement, and merits the name 'spiritual ecumenism;'"
- Fraternal knowledge of each other;
- Ecumenical formation of the faithful and especially of priests;

- Dialogue among theologians and meetings among Christians of the different churches and communities;
- Collaboration among Christians in various areas of service to mankind. "Human service" is the idiomatic phrase (CCC 821).

Within this text is the key to ecumenism in the media.

"A permanent renewal of the Church in greater fidelity to her vocation." The vocation of the Church is to lead souls to salvation by the announcement of the good news and the discipleship of others. Our interactions and exchanges online ought to be marked by this thoroughly. We can achieve this by ensuring that the words and example of Christ are present in in our text, voice, and videos.

"Conversion of the heart as the faithful." To convert is to become conformed to Christ. We can do this in the media by continually associating ourselves with peers who are like-minded in their continual conversion and with consumer media that is altogether rounded by the Scriptures and Church teaching.

"Prayer in common." There is a never-ending supply of friends and enemies for whom to pray. We must encourage our friends, family, and neighbors with our prayers and words.

"Fraternal knowledge of each other." In our media communities, we have a chance to interact on levels that are highly personal despite the distance and interface. Through this interaction and conversation, we ought to get to know those to whom we become close. This "fraternal knowledge" enables us to act as brothers and sisters as we engage the world in evangelization.

"Ecumenical formation of the faithful and especially of priests." There are many priests online. There are many religious online as well. Our lifting up of these people as our leaders is paramount to our ecumenical interactions online. This formation also means that we intertwine ourselves with our separated brethren as family, rather than as opponents.

"Dialogue among theologians and meetings among Christians of the different churches and communities." We should seek council and education from those who are set-apart in their scholarship, while, if given the opportunity and ambition, seem to emulate these great thinkers and defenders of the faith. We have to remember that each generation must replace the work of the theologians and apologists of the previous one. So get to work, right?

"Collaboration among Christians in various areas of service to mankind." Media that serves mothers, young men, priests, educators, and a long list of other occupations and interests can foster ecumenism among contributors and readers.

With the guidance of these seven ecumenical keys from the *Catechism*, the rest of this chapter will explore some of the practical opportunities that exist for those looking to be ecumenical with the new media as consumers or as producers.

Dedicated Websites

As I was transitioning in my faith from Protestant to Catholic, I became weary of anything on a non-Catholic website. For good reason, too! A meaningful part of my conversion was the realization that there were so many different denominations out there, and each one had a particular view on different aspects of the Christian faith and life. Some articles tinge on the foundation of *sola fide* (faith only) and others are outspokenly non-Catholic. I didn't want to chance any slant in this direction or that, so I chose an isolationist approach that would have me read only trusted Catholic content. I did that for a time and for me, it was healthy. Over time, I eased my way back into regularly viewing non-Catholic news sites, blogs, and other media. In that time of self-exile, I meant to focus specifically on Catholic teachings to start fresh on my faith and perspectives so I could have better clarity on the Church I was newly joining. I was decidedly Catholic even if I didn't know anything. And while there were things I was still in the process of learning, it was important to sequester myself from non-Catholic influences.

This was a smart choice. St. Paul, after his conversion, took a similar retreat from his former surroundings, spending three years in Arabia, and when returning to James and the rest of the Apostles, he did not lean into his own understanding or teachings (Gal. 1:17-19). Once we are ready, like Paul was, we ought to start dialoging with people of other religions through the new media.

Specific web domains have and should continue to be created that seek open, fair, and honest dialogue with those of other Christian persuasions. A great and recent example of this is *Strange Notions*, created by new media author Brandon Vogt. Brandon took his growing platform and used his network to form an online writing community that reached out to atheists for dialogue. Articles emerging from

Strange Notions have come from some of the top theist voices, and the atheist-to-Catholic converts have amassed a successful, appealing, and healthy following.

To be clear, ecumenical dialogue is dialogue with other Christians. *Strange Notions* is really a Catholic-to-atheist site. My point is that we should follow Brandon's example and create more sites dedicated to creating conversations to better communicate our faith with open doors and enthusiastic voices. Though many sites, podcasts, video channels, and radio shows already exist to produce great content to explain the faith, there is little in the way of conversing on the faith. There should be more initiative to bridge a digital link between Catholics and non-Catholics. Engagement is key. This sort of dialogue brings us together peacefully and intellectually. By such means, we grow closer to a unified Church, which is the hope of all Christendom.

These sites should not just be oriented toward religious affiliations but also toward other affiliations and interests like athletics, vocations, civics, social justice, and more. The possibilities are endless. As long as there's an important topic we agree on, we should work together.

If you have a blog or enjoy writing, and you are reading this, there is nothing stopping you from getting involved. There are thousands of young and older Catholics who must share their voices by their testimony, wisdom gained, and educational points to enhance others' faith. Why shouldn't you be one of them? There are virtually limitless platforms for you to create or to organize and join that can serve in this very effective mission and apostolate. This is how we change the world and bring the universal church closer to final unity. When better than the digital age, and where better than the digital continent, resident to millions who can access, read, and participate in ecumenical dialogue?

Bloggers Unite!

Who better to lead the charge for ecumenism than bloggers and columnists online? Often talking about news, doctrine, and life, bloggers are already well-equipped to carry on ecumenical dialogue in so many ways.

Each blogger should understand that while their page views might not exceed that of more popular sites, or more popular Catholic voices, they still have a unique voice and audience. In fact, bloggers possess an audience that those louder and well-known voices will never reach because they are familiar with and to their audience—they have actual friendships, not distant and virtual "friendships." Because we all have an aunt, brother, or loved one online, or someone we went to school with or used to know, we have an advantage over the saturation of social media.

Lay writers are specially equipped to carry out this work. Vatican II's *Apostolicam Actuositatem,* The Decree on the Apostolate of the Laity, offers the following:

> They exercise the apostolate in fact by their activity directed to the evangelization and sanctification of men and to the penetrating and perfecting of the temporal order through the spirit of the Gospel. In this way, their temporal activity openly bears witness to Christ and promotes the salvation of men. Since the laity, in accordance with their state of life, live in the midst of the world and its concerns, they are called by God to exercise their apostolate in the world like leaven, with the ardor of the spirit of Christ.[53]

---

53 *Apostolicam Actuositatem*, 2, available online at http://www.vatican.va/archive/hist_councils/ii_vatican_coun

With that said, let's go over some tips on how to have a more effective blog.

Have quality! More significant than any popular podcast or published author, individual bloggers are syndicated not by popularity or paid broadcast, but by organic quality. This makes them especially strong in their ability to speak with an ecumenical voice because, as the salt and light of the earth, they are ground in the lives of and are the leaders who illuminate the path for others. When people notice errors in facts, typos, repeated material, the credibility of the blogger goes down bit by bit. Strive for quality. We'll discuss this more later in the book.

Bloggers can be especially ecumenical in *the topics they choose to write*. Topics like apologetics, pro-life, education, travel, and the Catholic life are very popular for writers and attractive for readers. Anyone who writes on these topics must speak with a voice of unity.

No blogger should choose otherwise, but sadly, there are those who do. They use their platform to write about topics like being against those who receive communion in their hands, use words like "hate" when talking about the doctrines of other Christian religions,

---

cil/documents/vat-ii_decree_19651118_apostolicam-actuositatem_en.html .

and host a diatribe rather than a conversation. This is unacceptable, and really, it's unproductive for the vineyard. Such prose has no place in the Church. As far as restraint can help, those in the media should examine their own conscience before attempting to remove the speck from another's eye. (See Matthew 18:15-17 for a concise and effective means of dealing with this disagreement.)

The language couldn't be more clear. All Christians are called to this standard of goodwill, and those with public voices are held to an even higher standard because they are the direct communicators of the Gospel and truth.

So how else can bloggers write with ecumenism in mind?

The first thing to do is to *understand the religion and topics you are reaching out to or discussing*. Nothing is worse than a misconception spoken as fact, and as Catholics, we should know what it is like to be on that side of the table. It seems it is the nature of the world to miscommunicate Catholic dogma or to exaggerate a particular scandal or to otherwise place a black cloud over the Church. If we know how frustrating it is to deal with, we ought not to perpetuate the same on others. For that reason, bloggers must speak with precision when communicating the beliefs of other religions, especially those who profess Christ, baptize with the proper formula, and believe in the Triune God.

Next, bloggers can *support each other's work*. Here's some ideas:

- Police each other's comment-boxes.
- Provide friendly feedback on content and prose.
- Defend each other.
- Share posts that offer a different angle on subjects.
- Suggest content when struck with "writers block."
- Get to know each other outside of the media/sites.
- Pray for and encourage each other.

There will be no greater witness to the world than the message that's delivered when we show love to each other. All bloggers, micro-bloggers, and podcasters can be ecumenical when they support your friends online. You can encourage them in their ambitions, and when they need your prayers, you can stand with them in support. All Catholics in the media should promote the work that truly epitomizes this activity and cultivate a community of Catholics that works towards this end.

Never at the Expense of Truth

Your media has to be imbued in orthodox content, period. Everyone is called to preach the gospel, but we should take extra precaution to not "lean on our own understanding" (Prov. 3:5), and, just the same, we should steer clear of communicating novelties to others. Theological, apologetic, and evangelistic communication and experimentation should never be done at the expense of truth.

Previously in this chapter, I mentioned the important encyclical from Pope Paul VI, *Ecclesiam Suam*. Throughout the document, the late pope makes pleas for unity and demonstrates how—in so many ways—the Catholic Church places unity in the forefront of its evangelical and pastoral mission, but he ensures that readers know that concessions of truth will never be made in pursuance of those undertakings.

> That we, who promote this reconciliation, should be regarded by many of our separated brothers as an obstacle to it, is a matter of deep distress to Us. The obstacle would seem to be the primacy of honor and jurisdiction which Christ bestowed on the Apostle Peter, and which we have inherited

> as his Successor.
>
> Are there not those who say that unity between the separated churches and the Catholic Church would be more easily achieved if the primacy of the Roman pontiff were done away with? We beg our separated brothers to consider the groundlessness of this opinion. Take away the sovereign pontiff and the Catholic Church would no longer be catholic. Moreover, without the supreme, effective, and authoritative pastoral office of Peter the unity of Christ's Church would collapse. It would be vain to look for other principles of unity in place of the true one established by Christ himself. As St. Jerome rightly observed: "There would be as many schisms in the Church as there are priests."[54]

The prophetic pope makes it clear that any progress made toward unity will never be made at the expense of truth. This is a critical lesson for everyone involved in the media. Whether it is those who create humorous memes, satirical articles, children's books, or other seemingly harmless genres, all forms of media created for

---

[54] Paul VI, *Ecclesiam Suam*, 110, available online at http://w2.vatican.va/content/paul-vi/en/encyclicals/documents/hf_p-vi_enc_06081964_ecclesiam.html .

Catholic audiences—and why stop there?—should be devoid of error. After composing such a vivid call for unity in Christ in *Ecclesiam Suam*, literarily pleading for souls to embrace this true mark of the Church, Paul VI knew that the truth—even if it is the "long pole in the tent"—will be forever protected.

And it's a biblical fact, really. Jesus told Peter that "the gates of Hades will never prevail against it" (Matt. 16:18) If the Church was entrusted with an apostolic means of the transmission of truth, then by that charge, Jesus is the guarantor that the Church will never fall into error. It is, therefore, our job as media folk to make sure that remains true in our practice and profession.

Unity is ultimately a road paved with common ideas, and while we're sharing our takes on the Catholic Faith and on how we live our personal faith, we should always consider this when creating new media: will this promote unity?

Chapter 6

Bloggers Only! No-Nonsense Tips

I can remember way back to 1995 when my father got a contractor out to our house to set up a huge box with numerous wires and cables coming from it.

"What's this, dad?"
"It's another phone and cable line in order for us to have an Internet connection."

I still had no idea what he was talking about, but I began then to notice the numerous commercials for America Online (AOL, for those who can't remember) and it looked tantalizing. Endless and easy access to shopping, games, information, email, chat rooms, travel booking, sports and entertainment, and much more. The first time my mom logged in and opened a page on Netscape Navigator—the 1996 equivalent of Google Chrome—we were thrilled. It took a few minutes to download pages back then—and do you remember the clunky, garbled, static sound of starting a dial-up connection?—but the possibilities were endless and accessing information at this rate was almost science fiction.

Back then you basically paid for Internet like you did a phone line, so us kids were not able to get on at any moment we wanted and had limited access when we did. Expecting an important phone call? Too bad. It used a phone line and if you were on the Internet, you couldn't make or receive calls. On the phone and someone tries to get online? Too bad. So when I was a kid, the Internet was more of a "mom and dad" thing. My mom would sit on her Gateway PC for some time and then would be off. When we could—and were allowed to—get on, we had to do it as a group, so we didn't do it too often.

Years went by and the Internet evolved. eBay had commercials during the Super Bowl, Amazon only sold books, networks were through cable, too, so browsing speeds increased, and there was a page for seemingly everything.

Fast forward fifteen years, and it's 2011. I had finished a short career in the military, married my sweetheart, graduated college, and deeply loved the Christian faith. I was highly interested in spirituality for a while, but my interests turned more intellectual as my faith was challenged more and more. My wife bought me the *Apologetics Bible* with a HCSB translation which was derived from the KJV. I was not yet Catholic at the time, but I had a deep love for the Bible and associated study. If there was a question, I wanted to answer it. If it was a matter of debate to people of high standing among Christians, I probably gathered an informed opinion, but left the debate there.

Then I got into blogging.

Blogging changed everything for me. I had no audience and no "followers." I'll be honest, I didn't really know what I was doing. At all. But if I was going to put content out there I wanted to make sure it was memorable, high quality, and worthwhile. If I remember correctly, I had five post views a month on my blogger.com page. Those views were, in all reality, likely from me refreshing the page after edits. Hence, I had no readers, and I wasn't really part of any community that shared or read my work. I discussed apologetics and current issues, but I didn't find any breakthrough niche. It was pretty much just me and my already-Christian friends. And to put things into context, they were all Evangelicals.

Soon I would convert to the Catholic Faith and things really picked up. I started a master's degree in dogmatic theology with Holy Apostles College and Seminary. A semester into the program and I had a lot to talk about and started to write for several Catholic websites to include *Ignitum Today, Catholic Exchange, Aleteia, Catholic Stand*, a number of others, and started my own site shaunmcafee.com where I could use any topic I wished. I did my best and got a bit competitive with myself. I admit, I enjoyed watching the stats of my post each day, seeing the reactions I could gather, and what folks might say in the combox. Writing for these sites enhanced my skillset for sharing, learning from others, gathering an audience, networking, and developing content.

A short time later, I had made a few inquiries to the vice president of our college, Sebastian Mahfood. Perhaps a little bluntly, I asked why the school did not do more to reach out to students, and why there was no media presence. "Why don't you go ahead and do that for us?" he remarked.

"Okay, fine, let's do this!" and I started the blog for the college's Dead Philosopher's Society and built a brand-new media presence for the college. Alongside this I continued to write elsewhere until I founded EpicPew.com and needed more time to devote to that.

Six months later everything is going fine and I start to pick up some attention. I got my first book deal with Sophia Institute Press for *Filling Our Father's House: What Converts Can Teach Us About Evangelization* which enjoyed some success. Catholic Answers noticed my podcast and asked me to guest host Catholic Answers Live, and other sites began to inquire about my availability to help them in promotions or networking. At this time, I decided to deploy to Afghanistan—a life-long career and personal goal as a military man, which meant I had to turn down several big opportunities. I put all of that in the Lord's hands, though, weighing and discerning the balance between my blogging goals, career and family goals, and also the tempting whisper of my inner blogger's ego. It was an easy decision, though, just a few tough opportunities I had to pass up.

Nine months later I'm back, and while I was deployed I had somehow found the time to write a short book on St. Robert Bellarmine. I renewed my commitment to EpicPew and my other writing interests. Time in the desert was busy, but I had plenty to think about and came up with a long list of book and article ideas. Oh, and I was also moving to Japan at the time . . . did I mention that? So, once again, I had to think about all those opportunities I passed up to go overseas. Amazingly, those opportunities seemed to come back in plenty. I was asked to write a weekly blog for the *National Catholic Register* which was a huge honor, got to host Catholic Answers Live a few times while they were in a host transition, began writing articles for their flagship publications, and got a side job as a teacher's assistant at Holy Apostles for their core courses in dogmatic theology.

All of this went over very well. I became the editor for *Catholic Answers Magazine Online* for a period of time, completed another book entitled *Reform Yourself! How to Pray, Find Peace, and Grow in Faith with the Saints of the Counter-Reformation*, still edited posts for EpicPew, and tried to pump out a weekly article for my site and the *Register*. I have learned—and will continue to learn—a great deal about blogging and writing in general.

Now that I've given you the quick version of my blogging story, I want to share with my fellow or aspiring bloggers—in no particular order— some of the highlights of what I've learned, what will make you better, and what will set you apart. It will perhaps make you a little happier as well.

Waiting for Perfect Conditions

Look, right away we need to get something straight: if you're waiting on the perfect conditions, you'll never get started. Someone shared this pithy proverb with me a while back, and it stuck. If you want to be a blogger but think you should wait for something else to happen, perhaps for your popularity and cache to be higher, to find some readers, for the kids to be older, for your job to take up less time, or for some perfect break in order to get started, you never will. Likewise, if you have a blog or even a book started but think the perfect words, time, and conditions will come later, they probably never will. Yes, writing is about quality and content, but it is mostly about time. If you never pull the trigger to the gun, the race can never get started.

I've experienced those dry times where I've become convinced that I have absolutely no good ideas or would rather take a nap or play some video games. Times when I just don't think anyone will read my writing. Things have also gone haywire in the middle of a project. When I founded EpicPew, so many things went wrong immediately: I didn't have the money to pay for the site someone was already building, and when it was launched, the site crashed on the third day. I had almost no time to do all the social media for promotion, and constantly wondered how it would all turn out. Things weren't perfect and I was already incredibly busy . . . but I did it anyway.

The doubts you have about writing and your "platform" are common, but you must think and be uncommon. Keep three things in mind: First, realize that *as a Christian, you are called to evangelize*. Take this dimension of your vocation seriously and let the graces from the sacraments move you. Next, *go for it*. There is no better time than right now to jot down a few words, share it, rinse and repeat. Third, *it takes time to get good*—stick it out and you'll see the quality increase and everything else will follow naturally. Any way you look at it, just get the process going. If you're waiting for perfect conditions, you'll never get started.

Having a Delicious Menu

When you go out to a restaurant you've never been, don't you look around at the place? You get a take for quality of preparedness of your food when you look at other dishes being served. You'll get a feel for the welcome of the place when you are first greeted by the host and your waiter. Then, you'll look at the menu and your hunger will either increase or you'll get bored.

The same is true of your blog. Sorry, but people do judge books by their cover, and blogs, perhaps, receive even more critical attention. Maybe the menu is bent, pages are stuck together, folded where they shouldn't be, and maybe several items aren't available or common items cannot be found, or the menu contains graffiti. As you notice these red flags pile up, you'll be disappointed. You might order something simple and never go back again. Just the same, if your blog screams "last decade," cannot be properly navigated, contains broken links or pages that won't load, or has an unmoderated comment section with personal attacks, people will automatically associate this with the quality of writing. Their overall experience will be bad, and they will be turned off just because of the difficulty of being on your site. Your media is just as much about experience as it is about content.

Keep it balanced, though. It's not all about aesthetics. You can have a delicious looking professional menu with promising descriptions of the food, but if it doesn't match when you bite and taste it, that's the food version of clickbait! You don't have to sink thousands into a site to get people to read, but make sure that you have quality, function, and design, or at least that one of these does not suffer because of the others.

Fame is Irrelevant

Ever watch *Star Trek: Voyager* in the 1990s or online these days? You might remember the conversation with Reg, an engineer, and Seven of Nine, a former Borg (which if you didn't know, is a bit like a Christian convert in comparison):

Reg: You're the one who people are most looking forward to seeing.

Seven of Nine: That's difficult to believe.

Reg: Why?

Seven of Nine: I was Borg.

Reg: Exactly. You *were* Borg, but you escaped. And despite incredible odds, you managed to reclaim your humanity. No one's ever done that before; you're famous!

Seven of Nine: Fame is irrelevant.

Reg: Maybe. But you've given hope to everyone who's ever lost somebody to the Borg. You've inspired millions.

Seven of Nine: If you think my participation in the "hoopla" would be beneficial to others, then I'll comply.

Even if you've never binge-watched the show or have no idea what Star Fleet is, you get the point. Fame is irrelevant when you're participating in something bigger than yourself. This is paramount for all Catholic authors to remember. As much as we deny it, it's nice to see a high view-count, read an email from a fan, see shares, re-tweets, and so on. What's even better is to know you're participating in the greatest movement in human history, and the most important undertaking in salvation history: evangelization. Keep a fresh dose of humility on you at all times and beware of the temptations that come with your successes (or even your failures). Do not make a golden calf out of your blog or your stats. Keep this in mind at all times: it's for God, not you. Any success you have is his. Fame is irrelevant.

Mailing It In

It's a saying the older generation will remember. When faced with the task of transporting an important object to a destination, some people will hand-carry the document because they want to personally ensure its well-being while others will just mail it in.

Today, it's a figure of speech for a lazy

person. In writing, if someone really demonstrates a lack of delivery, care, and overall quality, they've mailed it in. I've experienced this first-hand, and it can be pretty obvious, and pretty ugly.

Once upon a blog, I had an unenforced obligation to write once a week for a certain online magazine when some weeks I did not have time nor did I have the creative juices flowing to really write the sort of article I usually could—one that people will want to read. A time or two, I mailed it in. It happened a limited number of times, but those stick out in my mind as sore-thumb occasions in which I could have done better, or done nothing, but I chose to write something useless which might be worse than nothing—do you prefer silence or static white noise? You get the point. Arguably, the only one that remembers is me and that says a thing or two.

Mailing it in might also be the time I didn't proofread my own work before sending it to an editor. Rather than respect his time and where else his attention could be placed, he had to spend a significant amount of time correcting things I could have squared away. If I didn't care, why should he?

When you write, proofread. Are you looking for common errors, properly citing and linking to sources, and ensuring the correct translation of the Bible was used? Did you comply with the style guide you've been provided? Don't leave it to your editor to fix—it's *your* name that shows up on the post. Do you want people to know you for typos or great teaching? When you write, do your very best every time. Don't worry about having the exact words that will accumulate a zillion views and shares: just communicate succinctly and with passion. *Really care about your words and your quality!* Never mail it in.

Worth Reading Twice

Start to remember this one word every time you write: ephemeral. It means something is passing, momentary, short-lived. The majority of writing is ephemeral in general. The nature of online content is particularly short-lived. A post might hang on the "top viewed stories" for a bit of time, or it might even enjoy a time of viral viewership, but it eventually goes to the Island of Forgotten Articles.

But, some don't. Some articles can be re-shared, re-read. What makes these articles different? Firstly, the content is not fixed to a story that nobody remembers anymore. Any-one want to read about Mary Kay Letorneau?

Crickets. Second, some content is still relevant. This next Easter, I bet you'll read something about the Resurrection that was not recently published. Or that post about Santa that gets shared every year—yup, you'll see it again. Third, it was controversial and still makes a good point. One post that comes to mind is Mark Shea's criticism of the anniversary of the atomic bomb being dropped on Hiroshima. When I first read the post I was anxious to comment, only to realize the post was over a year old. I've seen it several times since. The same thing happened when I posted a ditty about the sex-fueled circus we call the Super Bowl Halftime Show. It went Catholic-viral some years ago and I've been able to re-share it with great success each year since. Lastly, some writing is just unforgettable. It seems that every Catholic convert has read *Rome Sweet Home* by Scott Hahn or *Catholicism and Fundamentalism* by Karl Keating. Many have read both, multiple times. And it's not because of some publisher's conspiracy to push these as the standard, or some marketing strategy to capitalize on the heart-strings of converts—it's because these books are extremely well-written, well-edited, and most importantly, well-remembered.

What's my point? It's not to say that commentary on recent news is a poor use of your time or that you shouldn't write on an idea that is important to you. But when you write, write something worth reading twice. Write something worth citing, worth quoting, worth searching for to share with a friend who brings up the topic in the future. Research, document, and tell a story, and be diligent about it. Doing this online is a struggle because there is an endless and undying stream of new content being written, published, shared, and eventually replaced by something new. Writing online is ephemeral, so your best to be unforgettable. Write something worth reading twice.

Mentor Someone

Those who won't share their knowledge are either wildly insecure or wildly selfish. Probably both. Knowledge of success in writing, promotion, marketing, sales, social media, and networking doesn't require a secret security clearance with the Department of State, so don't fall into the practice of acting like it does. If you have information that might help Catholic authors share a good message or their testimony, or information that might help them boost their reach, share it.

When I got my first book deal, one of the first things I did was tell everyone about the book, which was a good decision. I did my best not to brag, but I let people know about the good news and success, and what to expect in the pages. Then came the editing process. Then the publication. Then the promotion. In these phases of unfamiliarity, it sure was nice when my more experienced colleagues would tell me what to expect and how to make sure I did everything I could to benefit the publisher and increase the value of the book. There were a few people who aided me with recommend-dations for references and resources that help with style, and others who gave me a few pointers on quotes. Others shared their experiences with me about the inherencies in being a published author. For example, one friend told me that "The publisher knows their audience better than you. It's not just *your* book anymore—the publisher will make changes that they must in order to reach their audience and make their sales goals. You have to buy into this process, not just for being a "yes man" but for trusting in their success and mission."

This information was spot on and invaluably helpful. Every bit of it was correct and since then, I've noticed how exacerbated people can get when their content goes to a publisher only to get "chopped up" or they're shocked to find out that they're largely involved and responsible for promotion. In my own book to which I'm referring, *Filling Our Father's House*, an entire chapter was removed. Another was cut in half. Whole sections—sections which I thought were not only beneficial but integral to the book's message—were removed. I could have fought for these pages, but at what cost? Not being a flexible author to work with? Not showing buy-in to my publisher's and editor's decisions? Was I going to argue that I knew better on my *first book* what to say than in their five-hundredth book? No. Not only would it have potentially damaged the content and flow of the book, but it would have likely damaged my reputation as an author, too. We have to realize that these publishers ask themselves, "Why yes, he has great ideas, but he's resistive, opinionated to no end, and will only work if he gets what he wants." This kind of attitude is the one you definitely want to avoid.

All of this is to say that the knowledge others shared with me—because they cared and wanted to see me succeed—was essential to my success and the success of my message. If you have information, share it. Don't be afraid of someone taking that information and rising above you, selling more books, having a better website, or being liked more. (And yes, these are real reasons people keep knowledge to themselves.) Even those who are not concerned with the relative success of others still might refuse to share information with others because they're not in their inner-circle of friends. Or, perhaps they don't share because they won't receive any benefit from sharing said knowledge, as if they need to get something in return for any act of goodwill. Yes, this exists. Don't do it. Plain and simple. Always be anxious to share what you know.

Maria Monk

A bestselling book emerged in 1836 that claimed to expose the abuse of Canadian nuns by Catholic priests. *The Awful Disclosures of Maria Monk* was a hit in every way. It contained controversy, sparked real-life news stories of arrests and investigations of the people who might have been involved with the "Black Nuns," and became a go-to meme for abuse in the Catholic Church.

The problem was, it was a complete hoax.

The initial testimony was so persuasive and popular that the masses simply assumed it was true. *Why would someone publish such a chronicle if it were not true?* The damage was done, though. Folks left the Church in droves, believing the story which caused such astonishment and shock. It took years for people at the time to put together the pieces of the puzzle that was defrauding the Catholic Church, but by that time the image of the Church was smashed and trust had to be earned back over the course of decades.

In our time, people convince themselves that they are quick to distrust information they find in the Internet, but the fact is, times have not changed: only the medium has. So, a story emerges next week about what the pope said in an off-the-cuff interview, or a story emerges concerning the latest allegations against our clergy, and sensationalist writers will jump on the bandwagon immediately. Publications of the highest repute are all repeating the same story of scandal. A few days later, the actual quotes and allegations emerge—they were harmless—and the media made a such buzz about the event that nobody even cares what was actually said – they only care about the image the media and pundits were able to project.

Yes, there will be great Catholics out there who work to share the truth of the situation, whatever it might be, and it will fall on some attentive ears, but mostly deaf ones. And the bloggers and media folks who feel the need to say something are suddenly backpedaling to cover their tracks and salvage the situation. It's unfortunate. But it does happen all too often.

As a writer, then, what I've learned is to observe. Just observe. If something is news, it's okay to discuss, but it is wiser to not opine until all facts are discovered and made available to provide to readers. It's not just opining that's the problem in these situations, but it's also being quick to accuse, quick to make conclusions, etc.

What's worst is the group of folks out there who rush to be the *first* to say something. They appear informed, intellectual. But when that truth emerges, they end up looking pretty silly. Almost two hundred years later, in a time when technology should allow peer review to snuff out falsehoods instantly, we still have our modern-day Maria Monks.

What's best is to wait, observe, and then engage. If you get the impulse to say something, it's usually best to warn people of this very thing: do not believe everything you hear, don't make a hoopla, and wait for the facts—registered mail usually travels a bit slower.

The *Catechism of the Catholic Church* gracefully reminds us:

> The means of social communication (especially the mass media) can give rise to a certain passivity among users, making them less than vigilant consumers of what is said or shown. Users should practice moderation and discipline in their approach to the mass media. They will want to form enlightened and correct consciences the more easily to resist unwholesome influences (2496).

Chapter 7

Error 404: Avoid These Blunders!

Count the number of ways you might read a map incorrectly and you might become too dependent on your nifty car navigation system. You might get to your point of interest efficiently and you might never miss a turn, but you'll never know what you're missing on the way. When was the last time you found yourself studying a map to find a better route, rather than depending on pre-programmed instructions?

Today, many people follow the same programmed instructions given to them, and ignore the signs. They arrive at their destination, but they're not sure how they got there—or how to get back! They become dependent on their devices and the information hurled at them. They rarely venture outside of that digital world.

Now, count the number of ways you might live the Christian life incorrectly. You can miss out on sacraments by depending on your own wisdom. You may give harmful spiritual advice to a friend, you might cause your brother to stumble, and you might lead others into damning temptations. You can go for years in sin and never realize it. Or, perhaps you do realize it and just don't think anything can be done about it. As well, you can go on thinking that just because you're a good person—whatever that consists of—you're definitely going to heaven. You can go on thinking the duty to evangelize is for pastors, authors, and everyone but you. You might even become convinced that all persons should live their lives as they see fit, regardless of how it affects them or others because, hey, who are we to judge?

These are the directions the world is giving you every day. These are the messages you're receiving in that harmless show on which you love to binge. These are the undertones of the godless world around you. Become dependent on this navigation system, and you'll miss more than a new restaurant you never saw before—you'll miss out on the meaning of life and your own salvation. Not to mention, you'll drive others off the road of redemption, and cause others to miss critical turns in their own journey. But these are the directions the world is giving you every day.

Maps matter. Good directions matter. Critical thinking, even about the destinations and routes we've travelled and become comfortable with, matter. As we become more dependent on our devices and as we become evangelists, apologists, and attempt to deepen the faith of others through the new media, there are real potholes that must be avoided. In this chapter, based on my years of falling into some of these holes or witnessing others trip up, I've collected a few of these dangers that you need to become aware of, and avoid, in order to become effecttive evangelists, apologists, and loving neighbors online.

"To Her Own Loss"

We can accomplish a lot with the media. Every day people are changing lives by raising money and awareness for their relatives and friends with traumatic injuries, disabling conditions, and life threatening diseases. I've observed the last days of children with cancer be filled with joy by the effort of a community gathered for support on Facebook pages. Stories about communities banding together to defend an activist or business owner standing up for their rights are all over Reddit and Twitter. These activists and culture warriors use their social media accounts to plan and execute their mission. From anywhere in the world a student can learn a new language, earn a degree, apprentice to a trade, seek advice as a new mother, learn answers about their faith, any virtually any other topic. Social media is like a digital battleground, and at the same time becomes the digital church, forum, and classroom.

Like all tools that may be used for good, this wonderful tool can be used for ill. Complimentary to this idea is a statement from *Inter Mirifica* regarding the proper use of media for good, and at times for ill.

> The Church recognizes that these media, if properly utilized, can be of great service to mankind, since they greatly contribute to men's entertainment and instruction as well as to the spread and support of the Kingdom of God. The Church recognizes, too, that men can employ these media contrary to the plan of the Creator and to their own loss.[55]

---

[55] Vatican II, *Inter Mirifica*, 2, available online at http://www.vatican.va/archive/hist_councils/ii_vatican_coun

Anything regarded as "loss" to the Church is a loss to all her members. This notion is particularly important to those who use the media as an outlet of political, moral, and ecclesial discussion. When the media is used by Catholics or Catholic organizations there is a ripple effect that disturbs all the faithful. When a Tweet is sent by a popular Catholic politician that promotes an idea which is contrary to the teachings of the Catholic Church, the entire Church suffers. Suddenly, because *this person* said it, and *this person* is Catholic, any Catholics who stray from that idea are suddenly fighting a culture war. If a popular Catholic personality is caught up in a scandal, and media outlets disseminate any story that will sell ratings, individual Catholics can feel as though they are under siege.

We have to reasonably anticipate this. Jesus told us that if we follow Him, because the world hated Him first,[56] we will suffer persecution to some degree.[57] Notwithstanding this fact, we don't have to invite the negative attention. Not just that—we can be careful as to not invite attention by actions that shame the Church, and not speaking in disagreement with Her

cil/documents/vat-ii_decree_19631204_inter-mirifica_en.html .

[56] John 15:18

[57] 2 Tim. 3:12

teachings.

This lesson is deeply important to Catholics today. The entertainment we choose, our sources of news we share, people we follow on Twitter, "Like" on Facebook, connect with on Instagram, and interact with elsewhere, are all under constant watch by those who are not Catholic, and not Christian. Most of the people who do not share our faith are like an intelligence agent and a little child. They're constantly monitoring us to see where we mess up, picking up on how we talk to see if they can trust us, and all the while wanting to be like us if we can walk the walk and talk the talk. When we do cut up, make mistakes, and act like hypocrites, an image of our faith is stamped in the minds of these outsiders. This is why we must be so careful of what we say and do!

It's paramount. People don't follow hypocrites. If we preach a message of unity, but constantly argue, we aren't doing it right. If we demand holiness in others, but post crude and offensive content, we're doing it wrong. If we call out others for their lack of catechesis, but fail to follow basic Catholic social and moral teaching, we're doing it wrong. If we blog week in and week out about how weak Catholic personalities, or even bishops are, tearing apart our own family and have nothing hopeful to say, we are doing it wrong.

The right response in each of every situation is *pause*. We should pause at every occasion to consider what we are doing and how it will affect others, both our non-Catholic and our Catholic brothers and sisters. Ask question: *will this cause scandal? Is this a balanced news report? Is this worth sharing? Who wrote this? Have I also checked other resources to be sure this is accurate? What is the actual context? What is the actual content up for measure?*

One of the best and most regrettably plentiful examples is the media that is created and shared regarding Pope Francis. Like no other before him, and no other in this time, his words are almost universally twisted by many in the media. His famous "who am I to judge" statement counts, still, as a binding mark in the minds of innumerable Catholics and non-Catholics around the world that he is in support of the homosexual lifestyle. But he has repeatedly held up the Catholic moral and social teaching on sexuality and parenthood. The same thing happens several times a year, but for other issues. I've been guilty, too, of confusing his words. In 2017, he made remarks about the goodness of the French translation of the Lord's Prayer, otherwise known as the *Our*

Father. He remarked that sometimes the translation, “lead us not into temptation” is deceiving, and could better be rendered as “don’t let me fall into temptation.” Now, I did not believe the news stories, as many did, that he was about to impose sweeping changes on a translation throughout the world. I did, however, not clarify this position or understanding of the news story. What I did was I wrote a thoughtful Facebook post about the topic, and how I thought nothing was wrong with the translation—even defending the translation. What I did was not wrong, or directly harmful, but without even realizing it, I sounded like I was responding to and verifying the news that he was about to change the prayer. Because of this, I caused some serious grief for my friends and acquaintances. Perhaps, too, I lost a little credibility that day.

Simple mistake, but important moment from which to learn. Pope Francis’ words can be confusing. Not just his, but the typical priest’s words can be confusing, and St. Thomas Aquinas’ words can be confusing. A simple *pause* would have helped the situation tremendously. The best thing we can do when we pause is to consider all the questions italicized above, and proceed in three directions:

1) Say what we mean to say, how we mean to say it;
2) Share something said by someone else that adequately covers the topic and sentiment; or
3) Be silent.

Number three, here and in many cases, is a most noble choice.

I leave this section with the wise words of the Council Fathers of Vatican II:

> The Church recognizes that these media, if properly utilized, can be of great service to mankind, since they greatly contribute to men's entertainment and instruction as well as to the spread and support of the Kingdom of God. The Church recognizes, too, that men can employ these media contrary to the plan of the Creator and to their own loss.[58]

Not the "ME"-dia

"There is no "me" in team!"

---

[58] Vatican II, *Inter Mirifica*, 2, available online at http://www.vatican.va/archive/hist_councils/ii_vatican_council/documents/vat-ii_decree_19631204_inter-mirifica_en.html .

"Umm... Yes there is."
"Ugh. You know what I mean!"

Even though we fill in our profile information for social media accounts, and the sites ask about our personal information, identity, tastes, history, experience, and multimedia, the media is not all about us.

It might be a profile that belongs to and describes you and who you are, but the media is not about you, or me. People might follow you, like you, connect with you, and friend you, but it's not about the individual. It's about community. It's about relationships. It's about personal encounters with each other.

Too often, people make the media about them. Sometimes it's the blogger, podcaster, or writer who writes about their interests, what they ate for dinner, what they're reading, and how others can pray for them. They see themselves as "connected" but they refuse to reach out, only expecting to be reached out to. They don't seek friendships; they only expect people to seek theirs. These people are void of community, but we all know that even Chuck Norris—as cool as he may be—is a big shot in the movies. The lone wolves don't really exist. The bloggers and mediaphiles who expect the world to follow them but never become followers themselves will never truly touch lives.

The big mistake people make here is self-promotion at the expense of evangelism. Self-promotion can wittingly and satisfyingly make its way into the media we produce. Pope Benedict XVI understood this danger and addressed it in the 42nd World Communications Day message of 2008:

> We must ask, therefore, whether it is wise to allow the instruments of social communication to be exploited for indiscriminate "self-promotion" or to end up in the hands of those who use them to manipulate consciences. Should it not be a priority to ensure that they remain at the service of the person and of the common good...?[59]

In reality, we should never be followed with more attention than the attention we are directing to the Lord. When we are followed in the media, when we become leaders and friends to large audiences, the things they should be following are not the promotion of our works, but the promotion of works that glorify God and lift up his name. Music, art,

---

[59] Benedict XVI, 42nd World Communications Day, 3, available online at https://w2.vatican.va/content/benedict-xvi/en/messages/communications/documents/hf_ben-xvi_mes_20080124_42nd-world-communications-day.html .

good writing, epic humor, tasteful commentary, and the like should all be communicated with the promotion of the only one worth promoting: Jesus.

Talent and Toil – Don't "Purge"

The opening statement to the landmark media encyclical, *Miranda Prorsus* by Pope Pius XII, gives a bold introduction to the user and producer of media.

> Our generation takes great pride in the remarkable products of its technology, but even though these advances are the result of talent and toil, they are still gifts from God, our Creator, from Whom all good works proceed."[60]

This "gifts from God" statement came as a shock to many upon its release. Still today, it is read by many with a bit of surprise. *The Church calling media a "gift"?* Many would have you and I believe that the media arts, and especially the new media, are tools of the Devil—that nothing but gossip, indecency, arrogance, and

---

[60] Pope Pius XII, *Miranda Prorsus*, 1, available online at http://w2.vatican.va/content/pius-xii/en/encyclicals/documents/hf_p-xii_enc_08091957_miranda-prorsus.html .

ignorance comes from the media. Some of my own professors in theology and philosophy aren't completely convinced of the ultimate good of which media is capable. I hope that by this chapter—no, book—we've proven that meme to be very false.

As any tool in this world, even a good thing can be used the wrong way. Where man finds an opportunity to sin, man will. But that doesn't make the *thing* itself bad. A baseball bat can be used by a six-year-old to create a summer of memories, or it can be used by an angry mob to inflict damage to property and harm to others.

This is an infectious idea and I've watched it penetrate groups and cause close friends of mine to abandon media. I've watched countless people take up a "media fast" for a period of time, relieving their life of this tool. Most often it happens during Lent, where many people choose to "give up media" as their form of self-mortification. Others will "black out" their devices and accounts frequently in order to purge themselves of an apparent addiction or self-inflicted harm.

I can understand this from several points of view. Firstly, there's the issue of sin. When the device is used for an act that feeds an addiction and/or amounts to mortal sin, such as gambling or pornography, the device has to go (it should be cut off regardless of the liturgical season). I can speak from personal experience as to the amount of temptation that any device that can access apps or browser windows itself can represent. At times, it's necessary to completely flee from the things that aid in our sin and temptation.

Secondly, a person can abuse the device or social media account itself. As discussed, anything can be used the wrong way. Food, sex, freedom of speech, and money are good things, but they can be used to one's own loss when they are abused. Social media in particular can be abused when it is used in a disorderly way. That disorderly way might cover the paragraph above. Disorderly use might be counted as bullying, gossip, spying, etc. In many cases simply spending too much time on the device or account is the disorderly act.

In April, 2014, a video about this problem went viral on the Internet. Entitled "Look Up," this spoken word video created by Gary Turk challenged people everywhere to place their time and attention to something other than their phone's screen or risk missing the beautiful world around them.[61]

These are each great examples of when

---

[61] Gary Turk, *"Look Up,"* https://youtu.be/Z7dLU6fk9QY .

reducing use of media and media devices is smart, even necessary. When something comes between you, your God, and your family, you cannot achieve your highest of the "good works" the quote from *Miranda Prorsus* has inferred. Even if it isn't strictly getting in the way, sacrificing something we enjoy, for a time, can be spiritually rewarding.

I stated that I can understand the purpose of the purge in some cases. When considering those situations, a purge can serve as a worthwhile solution. There are even times when it might be laudable. A purge would be praiseworthy when on a retreat. When reflecting and meditating for a short period or for an extended time, it would be good to clear one's mind of the social noise. Rather than waking up and immediately checking for updates and notifications, a few moments with God would be more beneficial.

Again, there are many reasons to reduce the use of social media, but there's some context in need of discussion to finish this important lesson. I'm a Lay Dominican. It does me no good to avoid all the avenues possible for servicing, teaching, and preaching. The media are most certainly among those avenues. Severing this line of communication would be terminal on my ability to evangelize, which is the context of this chapter. Those who want to evangelize today's culture and witness the Kerygma to individuals must embrace the forms of communication available and used by this culture and these individuals.

Persons who purge themselves of their social media accounts might be doing so for a good cause, but causes being what they will, even the best of causes comes with a cost. If my favorite Catholic friends, the ones who inspire me, motivate others, and can be confided in suddenly remove themselves from my life during a time such as Lent or Advent, that has an effect on me and the rest of the social media community.

The world of social media is not build upon fake profiles and artificial intelligence. Real people engage each other in humor, debate, confessions, aid, fellowship, and growth. These are the same lives and the same personalities present in parish and community functions where they are physically present. Who would leave their parish community during its highest time of observance and celebration?

Social media is among the best means of evangelizing today. I respect the choice of others, but I consider the "media purge" or "media fast" to be extremely counter-intuitive to this effort. If the media arts are truly a 'gift from God,' then why would we not use them to the fullest and most correct extent in the times of joy, charity, celebration, almsgiving, and reflection?

As with all forms of mortification, even this seemingly small choice should be considered carefully. The fact is, God puts us in people's lives for reasons we cannot comprehend or understand, and if we are to take that matter into our own hands by removing ourselves from these relationships, we cannot possibly know the good we were chosen to perform.

Our highest commandment is to love God with all our heart, soul, and mind. Loving our neighbor is alike to this. Matthew the Evangelist tells us,

> When the Pharisees heard that he had silenced the Sadducees, they gathered together, and one of them [a scholar of the law] tested him by asking, "Teacher, which commandment in the law is the greatest?" He said to him, "You shall love the Lord, your God, with all your heart, with all your soul, and with all your mind. This is the greatest and the first commandment. The second is like it: You shall love your neighbor as yourself. The whole law and the prophets depend on these two command-

ments."[62]

The story of the Good Samaritan in Luke 10 is similar to this, only we receive a question: *"Who is my neighbor?"* And Jesus explains. In the digital age, if you ask, *"who is my neighbor,"* look no further than your friends list, your followers, connections, viewers, and contact list. Make the most of every 'gift from God' in order to bring God's kingdom to earth and his relentless love to the lives of others.

Stat Chasing

Stats matter. Ratings matter. The likes, shares, views, clicks, votes, tweets, and trending topics all matter. But they are not superior to the mission: the conversion of souls.

Many in the media will seemingly do just about anything in order to get a few more views or whatever media-metric they're seeking more of. Even when it comes to the Catholic media, bloggers flock to their keyboard to be among the first to provide commentary on the latest scandal or freak-out worthy news of the day. It's frightening. People with literally no formal education taking to the digital streets of Rome to tell other Catholics how to organize their

---

[62] Matthew 22:34-40.

thoughts and live their lives. And most of it is not positive because drama always elicits more emotion, causing people to want to share the content with others. That's one outcome of the new media: *everyone is suddenly an expert*.

There's also the other sort of stat chaser, the one who must check their site for traffic several times a day, no matter the content. This person is not satisfied with their contribution to the Church unless it is statistically measured in some way. I think we've all been there, those of us in the media. But what metric do you use to measure the impact of evangelization to a soul? How do you show value for the work of the Holy Spirit? Do you use Holy Points? Spirit Units? Bless-o-meters?

There is no way to measure this, so we shouldn't adopt or develop any sort of attitude like this. I did say views matter. The number of people to whom Peter witnessed after Pentecost was recorded for this reason: to show the impact of the power of the Gospel.[63] The key: that's not our power, it's God's. That means we cannot brag when 50,000 share an article we wrote, a video we produced, or a podcast we recorded. Even internally, we should cut the ego trip.

When I was growing up as a Southern

[63] See Acts 2:14-42.

Baptist, things always seemed to revolve around how many souls I would help save. *My, oh my, the pressure!* There is nothing so awful as to guilt someone into doing the work of Christ. And that's what stat chasing does; it depresses us until we reach some measure of success we were never meant to be measured by. Stats are ephemeral. They are a measurement of things that have happened. The real successes in media are the individual relationships we develop, and the lives that are brought to the perfection of Christ. One single convert, one single person helped from an addiction or suicide attempt, one soul brought back to confession, is invaluable to view. And if your media is monetized, consider the money changers in the temple![64] As my great friend and fellow Catholic author TJ Burdick will say, even if you get no views, Jesus wanted it all for himself.

Gossip

The new media allows for several different kinds of information to be procured, analyzed, assimilated, interpreted, used and archived. Education, entertainment, news and updates,

---

[64] I am not saying monetizing a site is a bad thing, just remember the principle of what you're really writing for: the conversion of souls to Christ.

religious services, are among these. There's one use within the media world that can destroy, impeach, and encroach on the lives of others. We all know that nagging and constant temptation: gossip.

Gossip has been defined in several ways. My favorite of all of these is what my pastor once said, "If you're not solving the problem, but you're spreading the problem, you're the problem." Gossip is the little remark we make that seems unbiased, that brings slander to another. It's the rumor. It's private information we don't have the permission to share. Gossip is a parasite that spreads quickly, disables vision, clouds judgment, and renders its host incapable of discernment and prudence. There might even be truth behind gossip, but it's usually not our information to share with people who don't need to know. In the military, it's called "need to know" information, and I think that's the best way we can think of it. Most personal information is *not* need to know.

With the media, there's some stand-out examples from which we can learn, some clarifications we can make, and some things we should avoid. Examples: scandals, "open letters," passive aggression, etc.

I would say there is a clear difference between libel and gossip. One carries the intention of destruction, and, for the other, destruction is consequential. The point of gossip is to spread privileged information, but pain to the victim is a result. Many of us never mean to hurt, but in the interest of curiosity, amusement, and conversation, we gossip.

Being part of the media for a bit now, I've come across people who have had trouble in the past. I've become friends with people who dislike each other. It starts with something small and turns into *"I can't stand that person. This person is a liar. This person cannot be trusted. Guess what she did? Guess what her husband did to me. You wouldn't be their friend if you only knew . . ."* I've heard the lot and always do my best to stop the conversation before it reaches a destructive course, and I do my best to be a peacemaker.

The root of gossip, I've already touched on, but the consequences of gossip can be deadly. In World War II, one of the most common problems was the handling of classified information. The qualification, arrangement, handling, and safe-keeping of classified information has not yet developed as fully as it did now. The complexity of the war added to the problem. The soldiers who were to storm the beaches in Northern France and the paratroopers ahead of them weren't even told precisely where they were being dropped, and they did not know when they were going until just hours prior. Many brave men who manned submarines had no idea where they were located. The phrase "Loose lips sink ships" gained popularity for its vivid depiction of the consequences of careless conversations.

Gossip, even in a more socialized context, is deadly. To such a degree, gossip can become a mortal sin. When our 'loose lips' release damning information about others, it can be cause for grave circumstance.

The *Catechism of the Catholic Church* reminds us that,

> The eighth commandment forbids misrepresenting the truth in our relations with others. This moral prescription flows from the vocation of the holy people to bear witness to their God who is the truth and wills the truth. Offenses against the truth express by word or deed a refusal to commit oneself to moral uprightness: they are fundamental infidelities to God and, in this sense, they undermine the foundations of the covenant (2464).

If you will talk about others, you should do so with the avoidance of a few things. First, avoid being polemic. Anyone can detect a snarky tone and this invites slight. Next, avoid exaggerating details. If someone wronged someone else, avoid expanding the facts. If someone tends to act a certain way, inform, don't insult. Avoid sharing private and personal information. Secrets are simple to detect. If you're unsure, ask the person for permission. Odds are that person will come to respect and remember the courtesy. Of course, this is an incomplete list, but it's a strong start. All you must do is ask yourself if you would appreciate others saying or discussing the information if it pertained to you.

Regarding the new media and the eighth commandment, the *Catechism* is not silent.

> Within modern society the communications media play a major role in information, cultural promotion, and formation. This role is increasing, as a result of technological progress, the extent and diversity of the news transmitted, and the influence exercised on public opinion. The information provided by the media is at the service of the common good. Society has a right to information based on truth, freedom, justice, and solidarity:

> The proper exercise of this right demands that the content of the communication be true and—within the limits set by justice and charity—complete. Further, it should be communicated honestly and properly. This means that in the gathering and in the publication of news, the moral law and the legitimate rights and dignity of man should be upheld.
>
> It is necessary that all members of society meet the demands of justice and charity in this domain. They should help, through the means of social communication, in the formation and diffusion of sound public opinion. Solidarity is a consequence of genuine and right communication and the free circulation of ideas that further knowledge and respect for others (2493-2495).

I highly encourage you to read through the rest of the *Catechism* on the importance and implications of the eighth commandment. See paragraphs 2465-2513. If all else fails, just remember this pro tip: "Do unto others as you would have others do unto you."[65]

---

[65] Luke 6:32.

Scroll, Don't Troll

Common to all forms of media is the inevitable dealing with accounts that choose to impose hardship, bullying, argumentativeness, and indecency on others. These folks sow discord in comment boxes, status updates, and everywhere else. For this behavior, they're rightly given the name "troll." When I first heard this term used in about 2008, it was more closely associated with a person who says something in a forum to solicit an emotional response. That's putting it lightly. Trolls seek the annoyance and irritation of others through idiotic questions, harsh remarks, or otherwise off-topic remarks. Whatever it takes to set people off, they'll say it. Anyone who has scrolled comment boxes for some time has come across the person who is willing to say anything to get a response.

This topic is important to the discussion of evangelism because all interaction online and in the media should glorify God. Trolls have the ability to bring out some of our more suppressed behavior, often triggering visible and tacit anger that devalues the otherwise fruitful works to which we are contributing. If we reply to the people who hurt us with the same rudeness and slander, we've burned down the very house into which we're inviting others to move.

Identifying an Internet troll is quite easy, but dealing with one is another story. There is a way to respond to them, and there is a very simple means of directing them elsewhere. Self-restraint can be a requisite for both.

If you're going to reply, do so with the utmost charity. While not being a doormat, being charitable means you are inferring to them the respect that you would like for them to infer to you. Initiating this can be a breath of fresh air, or, things can get worse.

When things get worse, that is, when you realize that your efforts are not working, your kindness is not directing the conversation in a constructive direction, abandon ship! This is not to say that we cannot have endurance when dealing difficult people. We must do everything we can to be Christ to those who have no light and seek other's humiliation.

There does come such a point, though, where you have to evaluate this, and decide that it might be best just to let another have the last word and seek that person's maturity and conversion in prayer, asking guardian angels to intercede and for Christ to lend a hand. Trust this, it is never us who converts others, rather it is us working as instruments of God.

My first suggestion said: “If you’re going to reply.” That leads us to the final tip. The time-tested way to deal with trolls is to simply ignore them. The same username from the month prior who did nothing but launch verbal assaults on your work will only feed off of your frustration. They sense it. Just leave them alone, be peaceful, and eventually, they will go away. Remember to always treat others with kindness no matter how they treat you. Read the words of Christ:

> For if you love those who love you, what reward have you? Do not even the tax collectors do the same? (Matt. 5:46)

Another form of trolling in the media has to do with the sort of people with whom we interact regularly. There are times when people post or discuss a link about a controversial topic and chastise those who disagree with their view. Likewise, there are those who go to the accounts of those with whom they disagree in order to pick a fight with something said, about anything really. There is an unwritten rule in the media that if you post something online, to a public forum, you’re not allowed to get upset when people disagree with you.

Too often I have been subject to this

behavior from readers and have seen it happen to others. It can get ugly. Good bloggers and Facebook users inevitably have their own trolls. That happens. Some take it too far, maligning someone for a comment with which they disagree, a comment that might have had all tact and care, or "unfriending" them altogether, etc. This one really astonishes me, when someone unfriends someone for posting something with which they disagree, and its contents are perfectly licit.

Case in point: One time I posted an article from author Mark Shea, who is known to be rather direct in his writing style. A particular individual whom I have known for some time since my early days of blogging commented, "Mark Shea? Are you kidding me?" I replied, affirming that it was a post from Mark. While waiting for his reply, I clicked on this person's page, and I noticed that I now had the option to send him a "friend request." For posting decent Catholic content, no matter from whom it came, but that this person happened to not like, I completely lost a "friend."

We cannot do this! We cannot populate our lives with people who only agree with us! From time to time, it does us good to read the work of people with whom we share interests, even if we disagree with their prose, tone, or opinion.

Would anyone support a politician who had a cabinet of "yes men"? No. My point in all of this, is that we can sometimes become the troll, without realizing it, by devaluing anything that challenges our views. Therefore, I have a saying: scroll, don't troll. When you see something you disagree with, consider the value of your contribution to the topic. If it doesn't lift the discussion, keep scrolling.

The Other Kind of Purge

Building a community is a big part of being involved in the media as a consumer and a creator. Making friends, establishing connections, and gaining followers are each equally important for the blogger and the blog reader. The same goes for the radio host, the television host, the podcaster, and all the other forms of new and old media.

Building this community is integral. It's not just about the followership or the friendships, but it's also about finding people who support you, that you count on to keep you in check, and who have the tactful courage to tell you when you're wrong or need a new perspective.

Eventually, this community will involve people who disagree with us. It might be someone who is aggressive, has an opinion contrary to the truth, or is persuaded by a different religion altogether. When controversy arises, and it always does, it is important to resist the temptation to block them out of your life completely. I've done this, and I've seen it happen to others, and there's a reason I almost always regret it.

Why? Well, for several reasons.

It's common sense to purge yourself of things that cause you to sin, but for me, it was a simple case of irritation and a desire to dump some of the drama. What I realized, though, is that if someone else's attitude, behavior, or beliefs caused me to be unhappy, there was more than likely a problem with me, not with the other person.

I came to realize that if I were to change the minds of those around me, and truly love them, I wouldn't reject them, deleting them from my life.

> If you love only those who love you, what reward is there for that? Even corrupt tax collectors do that much. If you are kind only to your friends, how are you different from anyone else? Even pagans do that. But you

> are to be perfect, even as your Father in heaven is perfect.[66]

Those are the words of Jesus, and I do my best to take them seriously. A little caveat: use good judgment with the advice I'm giving. If someone is truly out to hurt you or your family, it's better to snuff out a small fire before the whole forest burns down. Likewise, don't put gas on this fire, and when snuffing it out, be forgiving and tactful. Otherwise, don't purge your accounts or your life of people. People change. Waiting to see that change is worth the little moments of frustration.

---

[66] Matthew 5:46-48.

Conclusion

Over the last several years, I've held a number of conversations with mentors, friends, and colleagues as to the usefulness of the media. There are a number of people who are not convinced that the media is a tool for good. They'll tell me that, more often than not, the media brings out the worst in people: jealousy and envy, bullying and passive aggression, sin and vice.

I always tell them that any tool can be misused. A tool is a tool, and what makes the tool useful is the user. The user must use the tool in three ways: according to its purpose, with intention, and with skill.

Catholics may use the media rightly and successfully if they employ these three rules. The media has a sure purpose, and that's communication with others. Communications accomplish three special ends: spreading ideas, forming trust, and cultivating relationships.

When we do these, and we apply intention, we can begin to undertake the will of God, which is to evangelize the world and spread the gospel. Being intentional with media—through any of the devices that enable it—involves a message of goodwill, peace, and love.

Finally, Catholics may properly and with holy intention strengthen their own skills in the use of their tools. Being nimble with social media accounts takes time and practice. An increase in a dedicated number of followers rarely happens overnight, and if it does, the newcomers are usually not there to stay. Skill involves creativity, repetition, and heart – creativity to communicate uniquely according to one's personal style, repetition to build good habits and longevity, and heart to provide meaning in a world desperately searching for truth.

The media is a powerful tool, capable of reaching and inspiring millions of people in a single call to action. It can also be misused or neglected. This book has laid out, according to the teaching body, known as the Magisterium, of the Church, numerous ways of using the media correctly along with the life lessons of this author's failures and successes. Use the documents with which you're now familiar to make your chosen media platform a success, and always encourage others to use the media properly, too.

Made in United States
Cleveland, OH
29 June 2026

39418403R00114